W0018353

Effects of Electromagnetic Radiation on Living Beings

The objective of this book is to show in detail how electromagnetic waves existing in the environment can affect the electrochemical currents present in the brains and bodies of living beings that serve to communicate with their internal organs as well as with other living beings. These electromagnetic waves are distributed intensively by current means of communication (television, cell phones, radar, medical equipment, electrical machines, electrical networks, etc.) and by the stars in the Universe. Such waves can affect in one way or another the electrochemical currents of living beings, which seem to be currently interpreted as sensations, hypnosis, telepathy, intuition, spells, mediumship, visions and precognition, as well as other less widespread forms, such as telekinesis, radiesthesia, clairvoyance, precognition and teleportation. Several world-famous examples of these possibilities are illustrated in the final chapter of this book.

Effects of Electromagnetic Radiation on Living Beings

Felix A. Farret

CRC Press
Taylor & Francis Group
Boca Raton London New York

CRC Press is an imprint of the
Taylor & Francis Group, an **informa** business

Designed cover image: Shutterstock

First edition published 2025
by CRC Press
2385 NW Executive Center Drive, Suite 320, Boca Raton FL 33431

and by CRC Press
4 Park Square, Milton Park, Abingdon, Oxon, OX14 4RN

CRC Press is an imprint of Taylor & Francis Group, LLC

© 2025 Felix A. Farret

ISBN: 9781032994178 (hbk)
ISBN: 9781032994161 (pbk)
ISBN: 9781003604037 (ebk)

DOI: 10.1201/9781003604037

Typeset in Times
by Newgen Publishing UK

Contents

Preface

The main objective of this book is to present a broad explanation of the influence of electromagnetic waves on people's extrasensory perception due to the frequencies to which they are exposed in their daily lives, regardless of their willingness. The term 'extrasensory' is understood here as all information received through frequencies outside those of signals that commonly affect, consciously or unconsciously, the sensory organs of the human body, mainly carrying information to the brain.

In this text, as much as possible, scientific explanations are provided, trying to unveil many beliefs that fit perfectly within what is technically known today as environmental radiation. Such frequencies are present in solar rays, winds, environmental noise, modern electromagnetic waves used for communication devices (cell phones, TV, radio, computers, measuring instruments, radars, advertising panels), human voices and many other sources. The most discussed receptors of this information are the five best-known senses: vision, hearing, touch, smell and taste.

The mental phenomena covered in more detail in this book refer to perceptions that are difficult to explain without interference from beliefs or religions. In reality, all these phenomena seem to be just sensitivities of the human brain to environmental frequencies, personal feelings accumulated throughout life and the electrochemical sensitivity of the human body composition, mainly the brain. These mental phenomena refer more directly to hypnosis, telepathy, intuition, spells, mediumship, radiesthesia, visions and precognition, as well as other, less widespread, forms such as telekinesis, clairvoyance, precognition and teleportation.

This book is not intended to be in favor of or against any religion or belief. Its main objective is to try to establish scientifically how the brain can interpret the environment in which living beings live, aiming to answer why and how some physical or sensory phenomena have been interpreted throughout history as religious manifestations, supernatural powers, magic or beliefs. As there were no acceptable explanations in ancient times, people sought engaging, religious, belief-based and, sometimes, self-serving and biased interpretations. In many cases, these phenomena were even used as a deceptive device to convince more susceptible people of 'facts' that would advance the interests of others.

The author of this book asks the reader not to be intimidated by the technical information in the first chapters. However, deeper acceptance and understanding of the contents of the initial chapters are fundamental to accepting the arguments presented in the final chapters.

Felix A. Farret
(https://orcid.org/0000-0001-9310-7074)

Author's Biography

Felix A. Farret was awarded bachelor's and master's degrees in electrical engineering by the Federal University of Santa Maria, in 1972 and 1976, respectively, a specialist qualification in electronic instrumentation from the Osaka Prefectural Industrial Research Institute, Japan, in 1975, an MSc in harmonics in High Voltage Direct Current (HVDC) transmission lines from the University of Manchester Institute of Science and Technology (UMIST), UK, in 1981 and a PhD in electrical engineering from Imperial College London, UK, in 1984. He has been a member of Imperial College London (electrical engineering) since 1984 and was awarded a post-doctorate in alternative energy sources at the Colorado School of Mines, USA, in 2003. He is also the author of several chapters in national and international books, mainly related to the use of renewable energies and alternative energy, power electronics and harmonics in power systems. He has been a speaker on dozens of occasions in different countries dealing with electromagnetic radiation, alternative energy sources and new directions for generating renewable energy. He is currently a volunteer professor at the Department of Electrical Energy Processing at the Federal University of Santa Maria, RS-Brazil, where he has worked in undergraduate and postgraduate teaching since 1974. His main activity has been guiding research theses and dissertations in a multidisciplinary educational environment related to alternative energy sources, industrial electronics, generation quality, electronics instrumentation, the efficiency of distributed generation and integration of alternative sources of electrical energy. In recent years, he has coordinated several scientific and technological projects with alternative energy sources that were transferred to Brazilian companies such as AES-South Energy Distributor, State Electric Energy Company, Hydroelectric Plant de Nova Palma, RGE Energy Distributor, CPFL and CCE Control Engineering Ltd. These projects are related to the integration of small and distinct sources of energy; voltage and speed control by load for induction generators; the development of energy systems with PEM fuel cell stacks; the minimization of harmonics in HVDC transmission systems; electromagnetism applied in electrical machines and communication systems; maximization of electrical equipment efficiency; photovoltaic power generation; and surface geothermal energy to optimize electrical energy consumption.

1 Means of Perception for Living Beings

1.1 INTRODUCTION

In the 21st century, much has been spoken, seen and heard about the constant tendency to adopt solutions or approaches based on artificial intelligence (AI). One of the factors behind this trend is that it appears to be an evolution or something already scientifically on the way to representing natural intelligence, characteristic of the human brain. However, in everyday life it is known that there is an influence of several other phenomena, such as feelings, beliefs, intuitions, instincts, deductions, observations, absurdities, imprecise descriptive language and counter-positive evidence that human beings use, unlike in computer programming. It is observed that this tendency occurs in both human and animal brains and bodies, domestic and wild. Relationships like these seem to have everything to do with vibrations in solids, liquids or gases in the Universe causing the differences perceived through the senses of living beings and are discussed in this and the subsequent chapters of this book. The basis for this are the five fundamental human senses: vision, hearing, smell, taste and touch.

1.2 WHAT ARE NATURAL FREQUENCIES?

Initially, to clarify what oscillations or vibrations are, it has to be said that they can be defined as a cyclical movement of matter characterized by their frequencies, the instants of their beginning and their intensity. To obtain an idea or an introductory example of what an oscillation (or wave) is, we can discuss the cyclical movement of a fan wave that causes the appearance of wave fronts in the air that propagate in the environment. These airwaves can move the skin, the eardrums of the human ear, the hair and other senses, which will follow the low frequency of movement in the air produced by the fan. In the case of the skin, this vibration should establish a differential level of vibrations per second with other touched parts of the body. In particular, in the case of generating a sound, it is known that it is necessary for the eardrums to oscillate at least something like 15 times per second, with a certain intensity. Therefore, the brain will interpret them as an auditory sensation. If it is a previously known sound and already recorded in the brain, the person will know what it is, which could be music, a warning, the proximity of an animal, a person's voice, thunder, an imminent danger, etc.

DOI: 10.1201/9781003604037-1

It is known from engineering, physics and chemistry that any material is made up of atoms that include electrons, protons and neutrons. Therefore, the constitution of a moving fan as well as the air around it, as mentioned in the example above, move atoms around. Therefore, the movement of electrons in the air can be considered as an electromagnetic movement or current and, therefore, capable of propagating electromagnetic effects. All this has to do with brain rhythms in animals, including humans, which are affected by fluctuations in electrical potentials generated by the depolarization of a number of local pyramidal neurons, also called postsynaptic potentials. These potentials are caused by electrical discharges of very low voltage (thousandths of volts or millivolts) that stimulate or inhibit other neurons that are in a harmonic and synchronous tune. When performing any task, there is not necessarily synchrony between these harmonic waves.

The sense organs and a network of nerves provide the necessary information to the brain through various organs in such a way that it can interpret and respond to the physical phenomenon that specifically generated the vibrations. Through these bodily senses, there is an association and interaction with the environment of the five best-known organs, namely, the eyes, ears, nose, mouth and skin. These five sense organs contain receptors that transmit incoming information to the appropriate locations within the nervous system via sensory neurons. Therefore, receptors for this information can be classified into two forms: general receptors, which are present throughout the body, and special receptors, which include chemoreceptors, photoreceptors and mechanoreceptors.

This chapter initially presents the main sensory phenomena of the human body as an introduction to the subjects of the next chapters.

1.3   WHAT CAUSES THE SENSATION OF IMAGE OR OPHTHALMOCEPTION

An image is the perception that human and animal brains have of a complex set of colors gathered in a flat or winding environment that reflect a certain color distribution under the different very high frequencies that make up the color vibrations thus generated. In the case of documentary or journalistic photographs, they are an instrument for the detailed recording of reality or are even used simply as a form of artistic expression. For these reasons, it is not possible to obtain a simple mathematical formula that is capable of expressing such complexity of frequencies.

Images can cause strange sensations for those who perceive them due to the complex distribution of their details and the way they are previously registered in the brain. For example, the colors in photographic images of nature are capable of bringing inexplicable sensations of comfort, memories, tranquility and joy, and can even reduce anxiety for those who observe them. An interesting related phenomenon is known by the term 'déjà vu' (in French, "already seen"), which was described in 1876 by Émile Boirac, French philosopher, researcher and president of the University of Grenoble in 1898. Psychoanalyst Sigmund Freud described déjà vu as the "memory of an unconscious fantasy combined with the desire to understand or improve the current situation". Carl Jung, in turn, theorized these sensations, stating that there was a relationship with the collective unconscious. James J. Giordano, professor of

Neurology at Georgetown University in Washington DC, USA, established that "there is nothing supernatural about this, and it is extremely normal for anyone to experience 'déjà vu'". It should be noted, however, that this sensation has been described for centuries in one way or another since Plato as if it were proof of the existence of past lives. Currently, many intellectuals try to explain this phenomenon as being another addition to the understanding of the human brain [1].

It seems that déjà vu is literally a subjective experience for people to try to repeat perhaps a certain desired set of events, activities, thoughts and feelings, even if it has never occurred to them before. Interestingly, the phenomenon of déjà vu diminishes in old age for approximately 90% of people who experience it as they become more experienced. This is still a mystery to science, which has understood it as something that the content of the imagination in the human brain creates and relates to the present with something similar to or different from what happened or was desired in the past or even with facts previously registered by the brain [2,3].

In recent decades, forms of image processing have emerged using computer hardware and software that manipulate digital data to produce images that can be modified or highlighted using specific information of interest to the author. As they are digital representations, these images are still a special representation of two- or three-dimensional scenes in the rows and columns of a mathematical matrix, or the pixels of this matrix. A pixel is the smallest unit that can contain individual color information. The numerical values in the pixel demonstration vary uniformly, ranging from 0, that is, no color (the black pixels), to 255 or even the content of all colors (the white pixels). Therefore, the more pixels an image has, the better the gradation of its definition.

The detail of a digital image depends on the number of pixels in its representative matrix, whose elements may be binary, octagonal or hexadecimal. In a binary image, only pixels '0' or '1' are used, meaning off or on, respectively. This is the case if someone wants to represent a monochromatic image, where only black [0,0], white [1,1] or a gray combination of them [1,0] or [0,1] will be present. With an 8-bit format, up to 256 different color tones can be represented. The middle number 127 is the gray one in this case. The 16-bit hexadecimal system contains many more color nuances, making it the highest format for representing a color. With 16 bits the red, green and blue colors that make up the format of the basic colors known as RGB ("Red", "Green" and "Blue") can be represented, as further detailed in Chapter 2 of this book.

1.4   WHAT CAUSES THE SENSATION OF SOUND

Sound is produced by vibrations in the air caused by some vibrational phenomenon that generates regions of compression and decompression in atmospheric gases that alternate periodically with the same frequency as the source that produces these vibrations. In some regions of the surrounding air the molecules are more concentrated, and in others they are more rarefied, which is caused by the compressions and rarefactions traveling through the air and constituting what is called a sound wave.

To understand how electrical signals are generated by sound waves and sent to the brain, it is important to understand the form of the ear. This sensory organ is made up of three main parts: cochlea, vestibule and semicircular canals. The cochlea is snail-shaped and has two chambers divided by a membrane. These chambers contain a fluid that vibrates with sound, thereby also vibrating the hairs on the membrane, producing pressure and decompression in its surroundings. The vestibule contains receptors that, together with the semicircular canals, help balance this composition. These semicircular canals form $90°$ angles to each other, allowing the brain to know the direction in which the head is moving. The canals are filled with fluid and have calcium crystals in their lining. The eighth cranial nerve, or the auditory nerve, connects the inner ear to the brain, sending auditory and balance information to the body.

The hearing process begins when the outer ear collects sound waves from the ear canal to the eardrum. This begins to vibrate, causing the bones in the middle ear, known as ossicles, to vibrate as well. The ossicles then perform a back-and-forth action, sending a wave in the fluid inside the inner ear and stimulating the hairs lining the membrane of the cochlea. The tiny hair cells that line the cochlea's membrane are called stereocilia. This back-and-forth action forms electrical pulses to the brain through the eighth cranial nerve, which translates them into sounds.

Sound waves need a medium to propagate and are commonly transmitted through air and other gaseous, liquid or solid molecular materials. Sound does not propagate in a vacuum due to the need for vibrating molecules that transmit their movement from one to the other. A wavelike movement of gas molecules characterized by intensity, frequency and speed of propagation, then, constitutes sound, which in physics is represented by:

$$v = \lambda \cdot f \tag{1.1}$$

where

v = sound wave propagation speed in meters per second (m/s);
λ = sound wavelength in meters;
f = vibration frequency in cycles per second or hertz.

The characteristic of sounds, known as timbre, is the quality that allows the ear to differentiate the harmonic content from sounds with the same pitch and intensity, but emitted with different characteristics. One can cite as an example the musical note with the lowest tone on the piano, La_0, which has a frequency f = 27.5 cycles per second, or hertz, as it is known in physics. The highest musical note is Do_8, with a frequency of f = 4,186 cycles per second. Therefore, each musical note played by different instruments or generated by different sources has a tonality or a characteristic sound constitution that is different from that of others according to its timbre. Therefore, timbre allows the identification of sounds created by the voices of people, animals and nature according to characteristic frequencies added to the fundamental tone and the number and intensities of the harmonics present in it.

The difference between a musical sound and any noise is the frequency with which it occurs. In particular, the level or pitch of musical sound is characterized by its intensity, level and timbre. Hearing quality is what allows the animal's ear to differentiate a frequency between the lowest sounds and the highest sounds, that is, the sound tone.

Sound propagates in the ambient medium, typically air, and can change according to the air density and pressure, which may differ depending on altitude. It is known, for example, that the famous recording studio in Colorado, Caribou Ranch, won several awards for recording albums and works between the 1970s and 1990s. Michael Jackson, Elton John and Chicago, among other celebrities, visited. They all said that the voices, the trilling of the instrument strings and the percussion were different and that the conversion of analog signals recorded on professional equipment was the best in the world. This was possibly due to the low pressure and low air density at Caribou Ranch's altitude.

1.5   WHAT CAUSES THE SENSATION OF SMELL

Smell or odor is the sensation produced in the olfactory organ (nose) by volatile particles present in the air close to it and emanated by emitting bodies. The nostrils clean, moisten and warm the air that enters them. However, the characteristics of this inhaled air are present in the volatile particles that define the type of smell for each phenomenon previously recorded in the brain. When inhaled, the air moves from the nose to the lungs and the particles that cause the smell sensation adhere to the olfactory epithelium, inside the nasal cavity. As odor particles are present in the air, when absorbed by the nose, they reach the olfactory epithelium, which is the tissue located on the roof of the nasal cavity filled with neurons. These neurons inform the brain, which immediately relates it to being a new experience or an experience that has already previously been recorded in it (see more details in Chapter 4).

Among the types of basic smells perceived by humans can be mentioned citrus, chemical, woody, mentholated, spicy, sweet, perfumed, popcorn and rotten odors. In fact, the human sense of smell is capable of sensing up to 10,000 different types of smells, and the human ability to distinguish them is increasingly reduced with age. These nuances of smell are also registered in the brain and their presence is detected in a different way associated with each person's previous experiences.

1.6   WHAT CAUSES THE SENSATION OF TASTE OR GUSTAOCEPTION

Taste or palate is the sense with which humans and animals in general distinguish the flavors of what is being put in the mouth. The taste organs in humans are located in the taste buds of the tongue, which are sensitive to four fundamental flavors: salty, sugary, bitter and acidic. In animals, the ability to sense taste is much more distinct than that of humans. For example, the taste organs in fish are external, through the so-called barbels; in butterflies and flies, they are located at the ends of the legs and so on. Dogs and cats feel practically the same flavors as humans, but each with greater or lesser intensity since they have a different number of taste buds than humans.

The distinction between flavors on the palate depends on the number of taste buds in each living being. In the case of dogs, for example, their sense of taste is less acute compared to humans because they have a smaller number of taste buds, around 1700, which allows them to distinguish only four flavors: sweet, salty, sour and bitter. Human beings have around 9000 taste buds and are therefore much more demanding when it comes to flavors. On the other hand, dogs have around 125 million aroma sensors and are capable of sensing more than 10 times as much as humans.

1.7   WHAT CAUSES THE SENSATION OF TOUCH OR TACTIOCEPTION

Heat is the thermal energy caused by the vibration of molecules of a living body, defining the differences in vibratory temperature that are also characteristic of other surrounding bodies. Therefore, differences in heat and pressure between the parts contacted by living bodies cause the sensation of touch. These vibrations flow from bodies with higher temperatures and pressures to bodies with lower temperatures and pressures.

The flow of energy between bodies with different temperatures and pressures occurs until an equilibrium is established between them, at which point the temperatures and pressures of the bodies that have exchanged differences become increasingly close. Initially this exchange is rapid, but it dampens exponentially as the temperatures and pressures of the surrounding bodies approach an equilibrium point.

Living bodies have touch-sensitive nerves in the fingertips, tongue, skin in general and in some intimate places, which transmit touch through parasympathetic connections that can even cause the release of hormones. This is the case, for example, with the hormone that gives a feeling of happiness, serotonin, or even the stress hormone, cortisol. With all this, touch is slow compared to other forms of perception, but is still very powerful.

The International System of Units (SI) stipulates that the unit to establish the caloric level of a body is the joule (J), but the unit that has been most traditionally used is the calorie (cal). A calorie corresponds to the amount of heat required for 1 gram (g) of water to have a temperature change from $14.5°C$ to $15.5°C$. The mathematical relationship between calories and joules is that 1 cal is approximately equivalent to 4.18 J.

The heat of a body is only called sensitive when it receives or gives up an amount of heat capable of generating sensitive variations in temperature without changing the aggregation state of its molecules. Equation 1.2 is generally used to determine the amount of sensible heat received or lost by a body.

$$\Delta Q = c \cdot m \cdot \Delta t \qquad\qquad 1.2$$

where

ΔQ = amount of sensible heat exchanged (cal);
c = specific heat (cal/g °C);
m = mass of the substance (g);
Δt = temperature variation in degrees centigrade (°C) or degrees kelvin (K).

The specific heat 'c' is a characteristic quantity of each substance that can be used to determine the amount of heat necessary to raise or reduce 1 g of its mass 'm' for a temperature difference of 1 °C. Temperature variations can be expressed in both °C and K, as these two scales have 100 identical intervals ranging from 0° to 100°, thus representing the same variations.

1.8   CREATIONS OF THE SENSES

Human beings are quite susceptible to illusions generated by their senses, despite having acquired innate capabilities and experiences to process such complex stimuli. Vision is the most accurate sense since it is formed by the highest frequencies and is prone to being confused with reality. Stories of optical illusions have existed since the beginning of humanity and in many different cultures, and have caused enchantment, fear, different interpretations and confusion for each person in one way or another, depending on the impressions they received throughout their lives even before birth.

The interpretations of the senses can come in different forms, from the creation of non-existent animals, such as ferocious tigers, enraged lions, frightening monsters, huge horses, ferocious snakes and ghosts, through to the imagining of huge moving forms, such as trucks, floating ships, aircraft or war tanks. In any case, they have always been part of the human imagination because they are visualizations, ideas or mirages for the rich or poor minds that created them.

1.8.1   Optical Illusions

There are several optical illusions that generally fall into one of the four best-known categories: ambiguous, distorting, paradox, or fiction. There are physiological optical illusions that make people misinterpret them; for example, a still image that appears to be moving or an image that appears to be of a certain color when it is not. These illusions are caused by physiological appearance in the way the eyes and brain interpret them, which can be specific, such as brightness, angles, colors, sizes or movements. In addition to these, there are cognitive illusions, which are equally fascinating because different people according to their previous life experiences can interpret them differently. The classic examples of this are the optical illusions used in psychological texts, where an image can be interpreted in very different ways. Based on these differences, somebody can obtain an idea of what is going on in the observer's head.

An interesting type of human visualization are mirages, which can be categorized as 'inferior' (meaning those at low altitude), 'superior' (meaning those at high altitude) and the 'Fata Morgana'. The Fata Morgana is a type of superior mirage made up of a series of vertically stacked images created in an unusual way, which change quickly. Depending on the colors, the sensation of mirages occurs from the deviation of the straight line of light rays by a phenomenon known as refraction (see the example shown in Figure 1.1). There are three different environments here: water, air and the glass, each one with its own refraction coefficient. This phenomenon can be explained based on the light that reaches the eyes, which is sent to the brain by electrical signals that are interpreted as objects with similar shapes to other images that in one way or another are already stored in the memory.

FIGURE 1.1   Pencil refracted between air, glass and water.

Because of its nature, the brain can only see rays of light that propagate in a straight line, but the deviation of this property of light along the path forms the imagined images. Such light deviations can occur when the rays pass through different densities of media, such as the refraction caused by an image passing through a glass lens, from colder to warmer air and between water and air, among others (see Figure 1.1 and more details in Chapter 4).

Visually, the senses can be easily deceived by imagination, that is, by what has previously been experienced by each person's human brain with more or less intensity, as discussed in Chapter 3. For example, illusory images, such as the desert mirages that occur in the atmosphere, are due to real optical characteristics and can even be portrayed. These mirages can be low or high. It is common on intensely hot days, for example, to see images of vehicles on the horizon on large highways reflected by a puddle of water caused by wet asphalt, creating a clear impression. However, when approaching the place of vision, one realizes that the highway is completely dry and that the vision has been deceived. This is an example of a low mirage, as it is at ground level. For a mirage to occur, the ground must be very hot and the surrounding air distinctly cooler. The layer above the ground will be heated and the light ray will be refracted as it travels from the layer of colder air to the layer of warmer air.

The human brain can also interpret images of landscapes as if they were in motion, since the air is rarely completely still, causing the vision to see them as trembling or deformed. Air movement occurs through both wind and hot air, with temperature differences giving the sensation of a moving mirage known as a low-type mirage.

High mirages are much rarer, but they are much more impressive. These mirages occur due to the temperature distribution of the coldest air layer close to the Earth's surface and the warmest air layer well above the surface. This phenomenon is typical in polar regions or in very cold waters and can rarely be seen elsewhere. The object seen in these optical illusions appears to be much higher than it really is. This typically happens on the surface of the sea or glaciers, where temperature differences are much more pronounced. Someone can see on the horizon, for example, a ship floating in the air, or it can appear to be much higher than it actually is due to the different light reflections caused by the variation in air densities on the surface of the ocean in Arctic waters. In these maritime mirages, inverted images of ships can be formed, which are not yet visible due to the Earth's curvature, but which are reflected in the atmosphere like a mirror. This is one of those cases occurring at sea for some objects that are actually beyond the horizon, but appearing to be at a shorter distance [4–7].

Finally, there are literal optical illusions created intentionally by well-known artists such as those of René Magritte that can boggle minds and are used as a form of entertainment, or art, or to test attention or credulity. There are also accidental optical illusions caused by natural things such as atmospheric phenomena; for example, lightning, flashes and clouds with similarities to or the position of an object at a certain time, size or setting. In addition to these, there can be illusions due to some physiological change in the ocular system of the person in question.

1.8.2 AUDITORY ILLUSIONS

After vision, another sense that is very prone to illusions is the auditory one. Auditory illusions are false perceptions of a real sound or an external stimulus that are equivalent to an optical illusion: the listener hears sounds that do not exist but are present in the stimulus or sounds that should not be possible due to the circumstances of how they were created. The way these compositions are confirmed is believed to be largely responsible for the inaccurate judgments people make when evaluating information, since humans typically interpret and remember information that best appeals to their own experiences, beliefs, or interests. One of these erroneous interpretations, known as illusion, is the category of auditory illusions [8].

The brain uses several senses simultaneously to process the information detected by them. Spatial types are processed with more detail and precision in vision than in hearing. Auditory illusions highlight areas where the human ear and brain differ from perfect audio receptors to organic survival tools. This shows that it is possible for a human being to hear something that is not there or to react to a sound that they supposedly heard. When someone is experiencing an auditory illusion, it is because their brain falsely interprets their surroundings, causing them to distort their perception of the world around them.

Many auditory illusions, particularly those of music or speech, result from hearing sound patterns that are expected by the person, even if they are heard incorrectly. This is due to the influence of human knowledge and experiences of many previously heard sounds that can serve as a basis for new artistic creations. In this sense, it is interesting that the auditory system avoids the auditory echo created by the perception of several sounds coming at the same time from different sources and spaces. The human auditory system then selects these sounds as coming from a single determined source. However, this does not prevent people from being fooled by such auditory illusions [9,10].

The sounds found in words are called embedded sounds, and these sounds are the cause of some auditory illusions. A person's sound perception of a word can be influenced by the way they see the movement of the speaker's mouth, especially for people with hearing impairments, even if what they hear has remained unchanged. For example, if someone is looking at two people saying, "I know" or "I've never seen them", the word they hear will be determined by the viewer's greatest interest. If these sounds are played in a loop, the listener may hear different words within the same sound. More particularly, some cases of embedded sounds are those of people with brain damage who are more susceptible to auditory illusions and certain words that may be more common to them [11]. This phenomenon is known scientifically as speech-language pathology. This study is related to speech, fluency, eating and swallowing. In fact, it involves all speech and language mechanisms, along with the therapeutic application of corrective and augmentative measures to help people who have speech disorders to pronounce and communicate better.

1.8.3   GUSTATORY ILLUSIONS OR GUSTAOCEPTIONS

The tongue helps us to perceive various tastes and flavors. Taste buds are the papillae on the tongue that help distinguish different flavors. The senses of smell and taste tend to work together. If people could not smell something, other types of smells could confuse them. The sense of taste is also known as gustaoception. The taste buds on the tongue contain chemoreceptors that function in a similar way to those in the nasal cavity. However, the chemoreceptors in the nose detect any type of smell if they are related to the four basic types of taste buds that can detect different types of flavors such as sweetness, acidity, bitterness and saltiness.

Gustatory illusions have to do with what is generally classified as 'taste'. Basically, it is a set of different sensations that are not only due to the taste qualities perceived by the tongue, but are also important when associated with the appearance, smell, texture and temperature of a meal. The discernment of a flavor happens through the nose and only then will the flavor be combined with the real smell with which the food flavor was produced. If the sense of smell is impaired, for example by a blocked nose or a strong previous smell, the taste perception will also be altered.

Many food companies have used the taste sense as a way to emphasize or modify the flavors of their products. Interestingly, using a different taste test for the same product or changing the color of a white wine or soft drink can confuse even professional tasters. Changing the color of sweet drinks can be enough for people to perceive a completely different flavor; for example, a drink with a lemon flavor but red in color. Previously recorded brain information from previously felt flavors is the main cause of these distortions.

A bitter pill, citrus fruit, sour or just slightly sweet grapes are often associated with strong emotions. People involved in these types of tests express in words states of intense pleasure or displeasure depending on how they receive them. Such a strong link connecting taste to emotion and impulse has to do with human evolution, as this was a sense that helped test food before consumption. A bitter or sour taste was an indication of inedible poisonous plants or rotten protein-rich foods. Sweet and savory flavors, on the other hand, used to be considered a sign of nutrient-dense foods, barring a few unexpected surprises.

Tasty broth-flavored dishes evoke pleasant emotions for most people. They are a sign that the food is rich in protein. This flavor was recognized as the fifth basic flavor, in addition to the other four best known: sweet, sour, bitter and salty. The discovery that there are sensory cells specifically for this fifth flavor was made by a Japanese researcher around 1910, which is why a Japanese term commonly used for this is 'umami', which in the Japanese language means 'salty'.

1.8.4 OLFACTORY ILLUSIONS

Many studies have been published addressing the olfactory sensations of human beings, such as those addressing olfactory illusions in early blind, late blind and sighted individuals. It has been hypothesized that part of the illusion may depend on verbal labels evoking visual mental images of the odor source that, in turn, modulate olfactory perception. Specifically, in these studies it was predicted that label advertisements would not have initial effects on blind participants, as they were the least susceptible to olfactory illusions or, at least, to the effect of the label's odor on the perception of satisfaction and intensity. Such characteristics should be smaller in relation to individuals who see well and those who are late blind. With this, it was predicted that late blind participants and those who see well would not differ in olfactory classifications. Furthermore, it was hypothesized that, regardless of visual status, odors with negative labels should be classified as least pleasant and be more intense than odors with neutral and positive labels.

Odors with neutral labels should be classified as less pleasant and intense than those with positive labels. This is used an exploratory approach to investigate the potential effects of odor-related labeling and blindness on familiarity ratings due to sparse prior research. As the ability to identify odors can potentially affect the degree of olfactory illusion, an odor identification test was also included. However, as most studies show that blindness does not result in enhanced olfactory abilities, no group differences in odor identification were expected.

Smell is the human sense with which perfumes have become a great commercial and sensory differentiator over time in the composition of women's clothing, in particular. Men seem to be more reserved when it comes to their sense of smell.

Finally, to control for the potential effects of demand characteristics, data have been analyzed depending on participants' understanding level of the study logic. These studies concluded that demonstrations could fail for several reasons that are similar to those that can cause visual or auditory illusions to fail. For example, illusions caused by strong prior cognitive polarization.

Other factors can also be included, such as olfactory illusions like those that happen to some people who can never merge the random points of Julesz stereograms. Random dot stereograms are a very new discovery, developed by Bela Julezs, during the 1970s. This is the study case for some people who have never experienced some of the illusions described in the literature. However, these random spots are consistently experienced by most observers and can create considerable surprise when the mechanism that produces them is explained.

1.8.5 TACTILE ILLUSIONS OR TACTIOCEPTIONS

The sense of physical sensation known as touch mostly comes through the skin. The skin is the largest organ in the human body related to the touch sense, which is also referred to as tactioception. The skin contains general receptors that can detect both moisture and touch, pain, pressure and temperature that are distributed throughout the skin. Skin receptors generate an impulse and, when activated, are taken to the spinal cord and then to recognition in the brain created by previous experiences [12,13].

There are several definitions of what a tactile illusion is that can include or exclude different types of phenomena. An extreme view is that all perceptual processes are illusions to some extent. Another view is that an illusion is a misperception, a view that has been the subject of debate. The idea that illusions are deceived senses is not quite satisfactory, even if these illusions can be purposefully employed for deception or camouflage. Note that different stimuli can produce the same perception. This is the case with color perception, since it is known that different spectrum contents can obtain the same color. Thus, a perceived color can always be wrong given a class of equivalent stimuli, and this is not an illusion. On the other hand, equivalent stimuli form a rich source of illusions.

Few tactile illusions compared to visual or even auditory illusions have been described in the scientific literature. This is due to the fact that many visual illusions can be created from simple materials, such as pencil and paper, or even occur simply by looking at a natural scene under certain inappropriate conditions, as shown in Figure 1.1. Furthermore, with the development of computers and the Internet, countless websites provide a huge repository of visual and auditory illusions. However, there is little easily accessible information about tactile illusions, with the exception of amusing versions of what could be a case of dipleesthesia (diple meaning traditional 'wind musical instrument' and aesthesis meaning 'sensation') that occurs when holding a pen between the lips while pulling the corners of the mouth diagonally. To demonstrate and study a tactile illusion it is often necessary to configure equipment that can create the appropriate conditions, which rarely arise naturally in an obvious way [14].

Generally, it is a complicated electromechanical challenge to design machines that can provide mechanical signals to people under computer control testing. In fact, commercially available force feedback devices are used. The limitations of these devices can cause stimuli to be generated with a lower quality than those provided by hardware devices. Such device limitations may be due to too much friction, insufficient rigidity, excessive weakness or heaviness, generation of mechanical noise and so on. Any of these defects can affect resistance or even eliminate the resulting perception. On the other hand, non-programmable mechanical devices can make these studies systematic, such as the search for detection limits, or controlled multimodal conditions that are cumbersome or impossible to configure, but they still have advantages. These illusions are surprising, robust, easy to demonstrate almost immediately, and can be created in more elaborate and controlled ways using robotic equipment. Furthermore, the existence of equipment that can eliminate the need to rely on the experimenter's manual dexterity is described in the literature. Among these are the 'comb illusion', the 'ridge/trough illusion', the 'curved plate illusion' and the 'bump/hole illusion' [15,16].

1.8.6   OTHER SOURCES OF PHYSICAL ILLUSION

In addition to the five sensory organs of the human body presented in the previous sections, there are two others that help people orient themselves in the world and that contribute to the creation of physical illusions. They are the vestibular system and proprioception. The vestibular system is what acts as a sensory system for the body as a whole and is responsible for transmitting information to the brain about body movements, head position and spatial orientation. This system is also involved in motor functions and helps maintain body posture, body balance, head and body stabilization during movements, and identification of the body's orientation and posture in relation to the environment. Therefore, the vestibular system is essential for the balanced maintenance of normal movements and balance in the human body [17].

Charles Spence, professor of experimental psychology at the University of Oxford, runs the Crossmodal Research Laboratory, which studies how the brain integrates information from the five human senses to produce a coherent impression of reality. He noted that often perception modes clearly influence one another on the path to becoming a conscious thought and that all of this largely depends on the types of organic molecules detected by the olfactory receptors in the nose.

Spence tried to understand whether the food taste could be similarly shaped by sound. In these studies, the food known as Pringles (a brand of chips or crisps), which has several flavors, was taken into account. The original comes solely with salt, but commercially it also comes with the flavors of onion and parsley, cheese and 'barbecue', among many others. In a test carried out by Spence on some people tasting identical chips, he observed that almost all volunteers under the influence of Pringles reported that they were different, saying that some came from cans that had been open for some time and that others were fresh. When Spence analyzed his results, he saw that Pringles that produced a louder, higher-pitched crunching noise were perceived as 15% cooler than softer-sounding chips. The experiment was the first to demonstrate successfully that food could taste different when it was simply experienced under different means of adding or subtracting sounds within environments with different sounds [18–22].

Another system, known as the proprioception system, is described in the bibliography as a conscious or unconscious perception of joint position that serves to help the body identify the muscles, joints and limbs located in three-dimensional space and the direction in which they are moving relative to the body. Walking, running or kicking without looking at the feet, balancing on one leg, touching the nose with closed eyes and feeling the surface on which is being stepped are some examples that explain proprioception system function [23,24].

REFERENCES

[1] F. Akram and J. Giordano, Research domain criteria as psychiatric nosology: Conceptual, practical and neuroethical implications, Cambridge Quarterly of Healthcare Ethics, 26(4), 592–601. https://doi.org/10.1017/S0963180117000 10X, 2017.

[2] F. Bartolomei, E.J. Barbeau, T. Nguyen, A. McGonigal, J. Régis, P. Chauvel, and F. Wendling, Rhinal-hippocampal interactions during déjà vu, Clinical Neurophysiology: Official Journal of the International Federation of Clinical Neurophysiology, 123(3), 489–495, PMID: 21924679, Mar 2012.

[3] S. Nigro, S.M. Cavalli, A. Cerasa, R. Riccelli, F. Fortunato, M.G. Bianco, I. Martino, C. Chiriaco, M.G. Vaccaro, A. Quattrone, A. Gambardella, and A. Labate, Functional activity changes in memory and emotional systems of healthy subjects with déjà vu, Epilepsy & Behavior: E&B, 97, 8–14, https://doi.org/10.1016/j.yebeh.2019.05.018, 2019.

[4] P.G. Hewitt and M.H. Gravina, Física Conceitual, 12th ed. Editora Bookman, ISBN-13: 978-8582603406, 2015.

[5] The Columbia Electronic Encyclopedia, 6th ed. Columbia University Press, 2023.

[6] G. Coggan with contributions from Abi Le Guilcher, Amelia Bamsey, 27 must-see optical illusions that will blow your mind, Creative Bloq, www.creativebloq.com/features/optical-illusions, 19 Sep 2024.

[7] D. Braine, BBC News, BBD Forecaster, 2023, Jul 11.

[8] B.L. Scott and R.A. Cole, Auditory illusions as caused by embedded sounds, The Journal of the Acoustical Society of America, 51(1A), 112. https://doi.org/10.1121/1.1981302, Bibcode: 1972ASAJ...51R.112S, 1972.

[9] Angelique Scharine and Tomasz Letowski, In book: Helmet-mounted displays: Sensation, perception and cognition issues, Chapter: 13, Publisher: U.S. Army Aeromedical Research Laboratory, Editors: C.E. Rash, M.B. Russo, T.R. Letowski, E.T. Schmeisser, DOI: 10.13140/2.1.3684.4804, Jan 2009..

[10] B. Kråkvik, F. Larøi, A.M. Kalhovde, K. Hugdahl, K. Kompus, Ø. Salvesen, T. C. Stiles, and E. Vedul-Kjelsås, Prevalence of auditory verbal hallucinations in a general population: A group comparison study, Scandinavian Journal of Psychology, 5, 508–515. https://doi.org/10.1111/sjop.12236, ISSN 0036-5564, PMC 4744794, PMID 26079977, Oct 2015.

[11] T. Fukutake and T. Hattori, Auditory illusions caused by a small lesion in the right medial geniculate body, Neurology, 51(5), 1469–1471. https://doi.org/10.1212/WNL.51.5.1469. ISSN 0028-3878. PMID 9818885, 1 Nov 1998.

[12] V. Hayward, A brief taxonomy of tactile illusions and demonstrations that can be done in a hardware store haptics laboratory, Centre for Intelligent Machines, McGill University, 3480 University Street, Montreal, H3A 2A7, Brain Research Bulletin, 75(6), 742–752, 2008.

[13] N. Menche (ed.) Biologie Anatomie Physiologie. Urban & Fischer/Elsevier, 2012.

[14] W. Pschyrembel, Klinisches Wörterbuch. De Gruyter, 2014.

[15] R. Schmidt, F. Lang, and M. Heckmann, Physiologie des Menschen: mit Pathophysiologie. Springer, 2011.

[16] R.S. Davidon and M.F. Cheng, Apparent distance in a horizontal plane with tactile kinesthetic stimuli, Quarterly Journal of Experimental Psychology, 16 277–281, 1964.

[17] G.J. Hyland, How Exposure to GSM & TETRA Base-station Radiation can Adversely Affect Humans. Department of Physics, International Institute of Biophysics University of Warwick, UK, May 2003.

[18] Massimiliano Zampini, Charles Spence, The role of auditory cues in modulating the perceived crispness and staleness of potato chips, Journal of Sensory Studies, 19(5), 347–363. https://doi.org/10.1111/j.1745-459x.2004.080403.x, 28 Feb 2005.

[19] N. Twilley, How packaging can make food more flavorful. Letter from Oxford, Accounting for Taste, www.newyorker.com/magazine/2015/11/02/accounting-for-taste, 02 Nov 2015.

[20] C. Spence, Relishing the taste of food, Nat Food, 4(4), pp 342–343. doi: 10.1038/s43016-023-00735-8. PMID: 37117540, Apr 2023.

[21] J. Youssef and C. Spence, Gastromotive dining: Using experiential multisensory dining to engage customers, International Journal of Gastronomy and Food Science, 31, 100686–100686, 2023.

[22] Oleszkiewicz A. et al., Effects of blindness and anosmia on auditory discrimination of temperature and carbonation of liquids, Food Quality and Preference, 107, 2023.

[23] U. Proske and S.C. Gandevia, The proprioceptive senses: Their roles in signaling body shape, body position and movement, and muscle force, Physiological Review, 92, 1651–1697. https://doi.org/10.1152/physrev.00048.2011, 01 Oct 2012.

[24] A.M. Kinser, M.W. Ramsey, H.S. O'Bryant, C.A. Ayres, W.A. Sands, and M.H. Stone, Vibration and stretching effects on flexibility and explosive strength in young gymnasts, Medicine and Science in Sports and Exercise, 40(1), 133–140, https://europepmc.org/article/med/18091012. https://doi.org/10.1249/mss.0b013e3181586b13, PMID: 18091012, 01 Jan 2008.

2 Frequencies and their Effects on Living Beings

2.1 INTRODUCTION

In the previous chapter, the sensory organs of humans and animals, which are necessary to communicate with each other or with the natural environment, were described. It has been observed that in each of their sensory organs all living beings depend on the reception of certain characteristic oscillating frequencies. In this chapter, these signal frequencies and how they can manifest themselves in the sensory organs are analyzed in more detail, taking into account the structure of bodies and their physical and biological properties, as well as the types of environmental signals previously recorded by such living beings. In this chapter, ambient frequency effects on living beings are discussed, from very low frequencies up to extremely high ones.

2.2 FREQUENCIES OF NATURE

It is known that human and animal sensors in general are sensitive to vibrations or oscillations in the environment known in mathematics as sinusoidal compositions. Three fundamental aspects characterize these vibrations or oscillations: amplitude, frequency and phase. Amplitude is the maximum variation in intensity that a given signal reaches. Imagine, for example, a sinusoid where its amplitude will be the peak value it can reach, its frequency will be the number of cycles repeated per second and its phase will be its beginning from a reference point. In physics, hertz (Hz) is designated as the unit of measurement that characterizes a frequency 'f' by the number of oscillations or vibrations per second generated by any physical phenomenon. These variations can also be represented by a rotation along a circle and having a characteristic angular value at a certain point. In this case, the positioning will be given in radians/second (rad/s). As rad is not a unit, but rather a proportion between two angles and is dimensionless, both rad and hertz have the dimension per second. The phase can only make a comparison between two sinusoids, adopting one of them as the measurement reference, referred to as being the fundamental and of the lowest frequency.

In oscillations that exist in nature, there is always a fundamental frequency that is generally added by a sum of other higher frequencies, thus causing what is known

 DOI: 10.1201/9781003604037-2

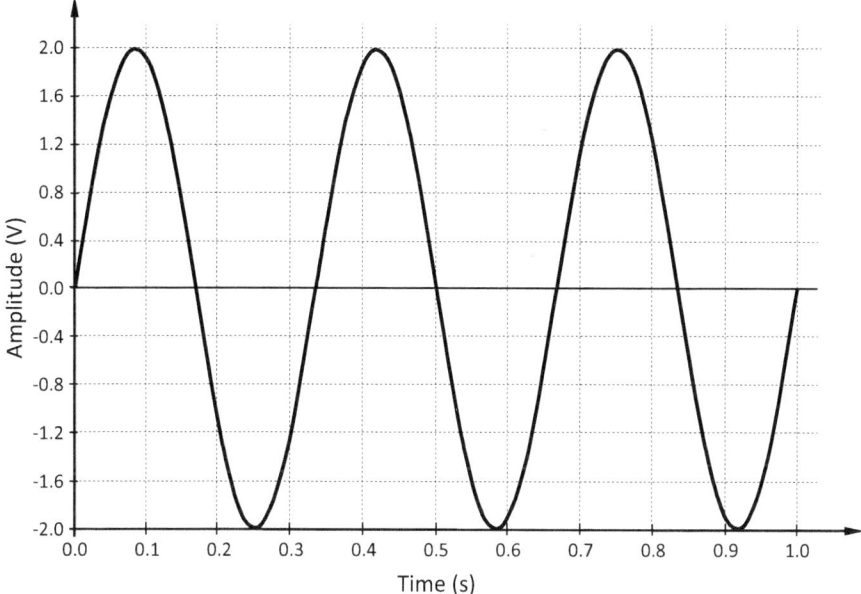

FIGURE 2.1 Parameters of a generic frequency.

as the sound timbre. As an example, Figure 2.1 shows an oscillation frequency that repeats three times over 1.0 second, that is, f = (3 cycles)/(1.0 second) = 3.0 Hz.

As amplitude 'A' is the peak that a repetitive signal reaches, the phase is the displacement from the beginning of the oscillation in relation to the reference used for the recording time. In the example in Figure 2.1, the phase starts at t = 0.0 seconds and the amplitude of the wave in this figure is 2.0 volts with a zero phase, since it starts at time zero. These are the mathematical bases used in this chapter.

2.3 CHARACTERIZATION OF VIBRATIONS DETECTED BY LIVING BEINGS

Animals, including humans, perceive environmental vibrations through their respective sensory organs, which can be mainly sensitive, auditory and visual. These organs are the skin, ears and eyes, which can be combined with the sensations of food taste and smell coming from beings and things in the surrounding environment.

2.3.1 BODILY SENSORY CAPACITY OF LIVING BEINGS

Skin receptors refer to specific sensations from corpuscles in contact with the body's sensing organs. These corpuscles, known as Meissner's corpuscles, are found in regions of the skin without hair and are specialized in the capture of tactile stimuli [1–3]. These receptors are formed by axons surrounding the hair follicle, which thus capture mechanical sensations applied against the hair. The endings or corpuscles known

TABLE 2.1
Skin receptors, stimuli and perceived sensations

Receiver name	Stimulus	Sensation
Meissner corpuscle	Rapid ringing vibration (20–40 Hz)	Rapid ringing vibration (20–40 Hz)
Hair follicle endings	Displacement by movement, direction	Displacement by movement, direction
Ruffini endings	Temperature, heat	Temperature, heat
Krause corpuscle	Pressure	Pressure
Pacinian corpuscle	Vibration (150–300 Hz)	Vibration (150–300 Hz)
Free endings	Intense mechanical, thermal and chemical stimuli; pain	Intense mechanical, thermal and chemical stimuli; pain
Merkel corpuscle	Stable indentation, touch, pressure	Stable indentation, touch, pressure

Sources: [1–3].

as Ruffini or bulbous corpuscles are one of the four mechanoreceptors or sensory receptors in the skin. They are mainly located in the reticular dermis of the fingertips and in the joints (see Table 2.1) [1–3]. The structure of Ruffini endings consists of dendritic fibers that branch into a capsule. Krause's corpuscles are structures formed by a nerve fiber, whose endings are shaped like a rod with one end thicker. These corpuscles are located both in regions bordering the skin with the mucous membrane and around the lips and genitals, capturing thermal sensations of cold.

Pacinian corpuscles capture vibrating and tactile stimuli. Free endings are sensory receptors linked to touch and located in the dermis of practically the entire body and branch out to the epidermis. Their function is to capture painful and temperature-related stimuli. Merkel's disks are nerve endings made up of axonal branches that end in flat expansions, related to the sensation of touch and pressure. The vibrational differences of different parts of the human body or external bodies stimulate the brain to perceive something that is already registered in its previous experiences.

There are some studies where the human body is considered as a single mass and the range of resonance frequencies is found according to the position of the body and the direction of vibration [4,5]. The human body will have completely different resonance frequencies for its internal organs depending on the direction and position of the person who is subjected to vibratory stimuli [6].

Several studies have been carried out with the aim of obtaining a dynamic sensory response for human organs. Some difficulties in these studies, however, were encountered when carrying out the tests, which can be attributed to the use of animals, mannequins or even cadavers instead of living people [4]. Therefore, the frequencies found in these studies may differ from reality, but they serve as an important source

TABLE 2.2
Resonance frequencies of the human body organs

Organs	Resonance freq. (Hz)
Abdominal	4–8
Shoulders	4–8
Lungs	4–8
Spinal column	8
Head	20–40
Chest wall	60
Hands and arms	20–70
Eyeball	60–90
Jaw	100–200
Skin	150–300

Sources: [4,5].

of reference and research. Such frequencies are nothing more than the transmission of contact movement from one cell to another. These frequencies can resonate, that is, one can directly influence the other depending on their dimensions and physical characteristics. Table 2.2 presents some resonance frequencies of human body organs collected from [5], as presented in [4]. For this, brain waves also have three basic characteristics: frequency, amplitude and phase. Each brain wave will be shaped by a frequency that is fired with a certain amplitude from the start of the ripple (phase zero). These components are responsible for coding (language) and decoding brain information, generating the most different waveforms. With this, each part of the brain will have greater or lesser activity with specific rhythms. These waves are known as beta waves in the motor cortex. In the visual cortex, they are alpha waves and in the hippocampus they are theta waves, as shown in Tables 2.2 and 2.3.

In addition to the best-known waves, Albert Einstein (1879–1955) also predicted that gravitational waves exist. Project researchers at the Laser Interferometer Gravitational-Wave Observatory (LIGO), based in Washington and Louisiana (USA), observed certain gravitational wave phenomena accompanying distortions in space with the interaction of two black holes 1.3 billion light-years from Earth. Mathematics can show how these waves can help unlock the mysteries of black holes, neutron stars, supernovae and other cosmic wonders that still defy human understanding. A singularity of black holes is the possibility that they can pull the entire Universe towards them. Therefore, the Universe and, consequently, living beings are disturbed by a gravity field that oscillates due to these interactions that have extremely low frequencies, nano (10^{-9}) or pico (10^{-12}) Hz. An experiment carried out at several measuring stations at different locations on Earth has already managed to synchronize the passage of a gravitational wave through the Earth. This phenomenon has probably led people to say that they are affected by the phases of the Moon and to believe that they can make predictions about events.

TABLE 2.3
Hearing frequency ranges of animals and humans

Wave frequencies	Frequency ranges (Hz)	Amplitude (μV)	Occurrence
Visual cortex Alfa (α)	8–12	3–50	Eyes closed relaxing and meditating
Beta (β)	12–28 (~30)	20	(a) Sensorimotor rhythms (SMR) of 13–15 Hz; (b) high beta waves, above 18 Hz; (c) low beta waves
Gamma (γ)	30–200	3 and 55	Related to concentration of thought
Delta (δ)	1–3	100–200	Related to deep sleep; certain pathologies are associated with them and generally occur in the temporal cortex
High theta (θ_a)	5.5–7	30 and 100	Recorded mainly in moments of intense emotion and deep meditation and related to hippocampal activities
Low theta (θ_b)	4–5.5	30 and 100	Also recorded in moments of intense emotion and deep meditation and related to hippocampal activities
Mu (μ)	7–11 (~12)	20	Spectrum of alpha wave located in the sensorimotor cortex related to upper limb movement

Source: [4–6].

2.3.2   HEARING CAPACITY OF LIVING BEINGS

The characteristic values obtained by Equation 1.1 shown in Chapter 1 for auditory vibrations can differ greatly from one individual to another. The reception of the propagation speed and amplitude of this ear ability will depend on the auditory and physical brain sensitivities of each person, also involving the timbre of the sound received.

As mentioned above, hearing and, therefore, the human brain, can distinguish or perceive sound movements in the air since there are at least something like 15 oscillations per second for a very low sound. That is up to 25 thousand oscillations per second for an extremely high-pitched sound. However, at birth, human beings hear much better than at any other life stage, but this is one of the senses that is most damaged over time, especially in people who are repeatedly exposed to very intense frequencies and noises. This is due to the joints of the internal bones and muscles that hold the eardrum, which become less and less flexible with age, making it increasingly difficult for sound signals to pass to the brain.

Whether a sound is heard and processed depends on its intensity, that is, the level of sound pressure it exerts when moving the tympanic membrane and its impact speed. This movement is transmitted to the hammer, anvil and stapes, which are tiny bones that move liquids from a bone structure called the cochlea, which differs slightly from individual to individual. In the cochlea are sensory hair cells that transform sound vibrations into nerve impulses. From there, these impulses are transmitted to the cortex by the vestibulocochlear nerve of the central auditory pathways. Based on this, an international standard for sounds was defined as ranging from 20 to 20,000 oscillations per second (Hz), which defines the standard range of sound oscillations covering most people. However, people who can go beyond these limits are very rare, usually musicians or people with high hearing sensitivity. In particular, Table 2.4 lists the hearing frequency ranges of animals, including humans, which are highlighted in bold for comparison purposes.

TABLE 2.4
Ranges of sound frequencies perceived by living beings

Animal	Audition range (Hz)
Pigeon	0.05–10,000
Mouse	1–150,000
Dog	15–40,000
Elephant	16–12,000
Bat	20–160,000
Spider	20–45,000
Humans	**20–20,000**
Cow	23–35,000
Cat	45–64,000
Lobe	30–45,000
Whale	40–80,000
Pork	42–40,500
Frog	50–10,000
Horse	55–33,500
Goat	78–37,000
Bird	100–15,000
Bottlenose dolphin	90–105,000
Chicken	125–2,000
Parakeet	200–8,500
Mouse	200–76,000
Chimpanzee	100–30,000
Sheep	100–30,000
Hen	125–4,000
Dolphin	150–150,000
Sparrow	200–18,000
Hedgehog	220–60,000
Rabbit	360–42,000
Sea lion	450–50,000
Shrimp	500–64,000
Boto	550–105,000
Orca	800–13,500
Seal	950–65,000
Beluga whale	1,000–123,000

Sources: [6–9].

2.3.3   VISUAL CAPACITY OF HUMANS

Similar to hearing, human eyes have the ability to perceive oscillations that typically range from 405 THz (1 THz = 10^{12} Hz) from the red color to the violet color that goes up to 790 THz (see Table 2.5). Some people can see more or less beyond such limits that are known as the visible spectrum for the vast majority of human beings. Infrared radiation falls below the number of vibrations of the color red and is known as thermal or heat vision. Ultraviolet radiation is that which goes beyond the vibration of the violet color and can hardly be observed by the eyes of the vast majority of ordinary people. These frequency ranges of color perception can be considered a specific quality or imperfection if the pattern of the auditory or visual brain system of each human being is considered [6–9].

The frequencies of smell and taste have much to do with the electrochemical reactions of human organic sensors (nose, tongue and mouth) to particles present in the environment or touch. These reactions cause the paths of the electrochemical current to change and the brain begins to perceive something that it had already stored through previous experiences. Table 2.5 brings together data on the highest frequencies that can be perceived by humans, similarly to other living beings in general, although at different levels.

According to the range of the electromagnetic spectrum that can be seen by the human eye, all the images that are seen are interpretations that the brain can produce about the electromagnetic waves that were emitted or reflected by the bodies and/or ambience around the person. The human eye is able to perceive these frequencies of light thanks to two special types of cells that line the back of the eye: cones and rods. Cones and rods are photoreceptor cells, that is, they are sensitive to perceiving light signals. Cones provide color vision, while rods are responsible for the perception of movement and the formation of black and white images such as those that someone tries to see in the dark.

There are only three types of cones in the human eye and each of them is capable of perceiving one of the following basic colors: red, green or blue (RGB). For physics, therefore, the colors that can be seen are nothing more than physiological phenomena that depend on the light capture and its brain interpretation. Furthermore, the proportion between each of the red, green and blue frequencies is capable of producing the other tones or colors perceptible by the eyes. When emitted together, these three colors produce white light, which is not exactly a color, but a superposition of visible frequencies. If the green light is removed from the white color, the pink color will appear [9]. Black is simply the total absence of light colors.

Normally, the basic primary colors of light are called RGB (red, green, blue), as they are additive, that is, the more colors that are added, the closer the result is to white. The colors RYB (red, yellow, blue) are the primary colors used in paintings, as they are subtractive, that is, the more colors that are added, and the result is closer to black.

In the retina, there are approximately 100 million photoreceptors to convert light into electrical impulses, but only the cones can detect colors. Rods help living beings see when there is little light. The retina uses a vitamin A derivative that absorbs light

TABLE 2.5
Limits of visual frequencies perceptible by humans

Color tone	Min. freq. (THz)	Max. freq. (THz)
Black	No color	
Infrared	<~405	
Red	~405	~480
Orange	~480	~510
Yellow	~510	~530
Green	~530	~600
Cyan	~600	~620
Blue	~620	~680
Violet	~680	~790
Ultraviolet		>790
White	All colors	

Sources: [7–9].

at night, and a lack of this nutrient can lead to night blindness. Electrical impulses with the color codes, luminosity and contours of the shapes of the observed objects travel electrochemically from the optic nerve to the brain. The cortex translates these impulses, detects movements and creates an image that is recorded in the brain. There is, however, a 'blind spot' in the connection between the optic nerve and retina perceived as a spot. The brain, however, compensates for this failure by interpreting it together with the image captured by the other eye. All this information is recorded in the brain, which helps the living being to interpret or confuse the newly recorded images.

The colors in ascending order of the visible spectrum perceived by humans are known as red, orange, yellow, green, cyan, blue and violet (Table 2.5). In addition to these, the color tones that humans can perceive depend on how much each of them contributes to the frequencies that go to the eye, including red, green and blue, to produce oscillations within each person's visual scale. Therefore, the colors and sounds perceived by human beings are nothing more than physiological phenomena that depend on the ability to capture light through the eyes and sounds through the ears and their respective interpretation by each person's brain.

Ultraviolet radiation (UV) is the band of non-ionizing radiation that lies close to ionizing radiation in the electromagnetic spectrum. The main natural source of this radiation is the Sun and its wavelengths can be classified in nanometers (nm) (1 nm = 10^{-9} m) as being: UVA (400–315 nm), UVB (315–280 nm) and UVC (280–100 nm). It must also be said that there are several sources of artificial UV, in addition to natural UV radiation, that can be found in occupational, medical and dental environments, including dental polymerization equipment, mercury vapor lamps, arc welding equipment and commercial UV bactericidal lamps (see Table 2.6).

TABLE 2.6
Ranges of ultraviolet light

Name	Abbreviation	Wavelength (nm)	Photon energy (eV)	Other names
Ultraviolet A	UVA	315–400	3.10–3.94	Long-wave, black light (not absorbed by ozone)
Ultraviolet B	UVB	280–315	3.94–4.43	Medium-wave (mostly absorbed by ozone)
Ultraviolet C	UVC	100–280	4.43–12.4	Short-wave (completely absorbed by ozone)
Near ultraviolet	NUV	300–400	3.10–4.13	Visible to fish, insects, birds, some mammals
Middle ultraviolet	MUV	200–300	4.13–6.20	–
Far ultraviolet	FUV	122–200	6.20–12.4	–
Hydrogen Lyman-alpha	H Lyman-α	121–122	10.16–10.25	Spectral line of hydrogen at 121.6 nm; ionizing at shorter wavelengths
Vacuum ultraviolet	VUV	10–200	6.20–124	Absorbed by oxygen, yet 150–200 nm can travel through nitrogen
Extreme ultraviolet	EUV	10–121	10.25–124	Actually is an ionizing radiation, although absorbed by the atmosphere

Sources: [10–13].

2.4 LIMITS OF HUMAN VISUAL CAPACITY

Infrared radiation (IR) goes beyond human beings' ordinary visual capacity and is released by bodies that emit heat, such as those coming from the Sun and fire. In addition to these radiations, humans can be exposed to artificial sources that include heating devices, lamps and infrared saunas. Laser beams emit one or more bands with extremely short wavelengths and are a special source of infrared radiation.

UV radiation also contains substantial energy, but it is absorbed in other regions of the eye before reaching the retina. All of this limits the human eye to a certain band of visible frequencies. Such diversity can be classified into different bands and categories of UV light, such as those described in the ISO standard ISO-21348 shown in Table 2.6. In addition, other common frequencies even higher than those are perceptible by humans. Among these are radio waves (3 kHz to 300 GHz) (1 GHz = 10^9 Hz), microwave (2450 MHz), X-rays (from 3×10^{16} Hz to 3×10^{19} Hz) and gamma rays (from 10^{19} Hz to 10^{24} Hz). As a rule, human senses cannot consciously perceive these frequencies because they are too high [10–13].

Light distortions may occur in nature as when a white light passes through an optical prism, producing colored beams of lesser or greater intensity because it has a density different from that of air. This occurs because the speed of the color waves varies with the frequency of each one. The different media densities through which light passes alter its propagation speed and distribution according to the frequency of each color. Therefore, colors acquire different refraction angles, each with its own characteristic frequency, giving rise to a continuous range of colors. Therein lies the explanation for the different colors assumed by the rainbow as it passes through air layers.

The phenomenon of environmental perception can be treated as a characteristic of each person's brain since not everyone has more or less the same ability to broadly perceive the spectrum of colors, sounds and/or heat around the human body. In short, the skin perceives lower frequencies, the ears perceive medium frequencies and the eyes perceive enormously higher frequencies. Table 2.6 is a list of the frequency ranges to which human beings are exposed, with secondary effects still to be better studied by science.

2.5   VISUAL COLOR CHANGES

Visual distortion depends greatly on the propagation medium and the receiver. Consider the white light that, when passing through an optical prism, produces colored beams of lesser or greater intensity because it has a different density than air. This occurs because the wave speed of each color varies with its frequency. Therefore, colors acquire different angles of refraction, as shown in Figures 2.2 and 2.3, each with its specific frequency, giving rise to a continuous range of colors.

The formation of an image on the retina of the human eye occurs through light that passes through the cornea and lens to generate visual stimulus, typical of visible radiation, the intensity of which reaching the retina will depend on the eye pupil. An analysis of different parts of the eye shows that the retina is very sensitive, as it has the

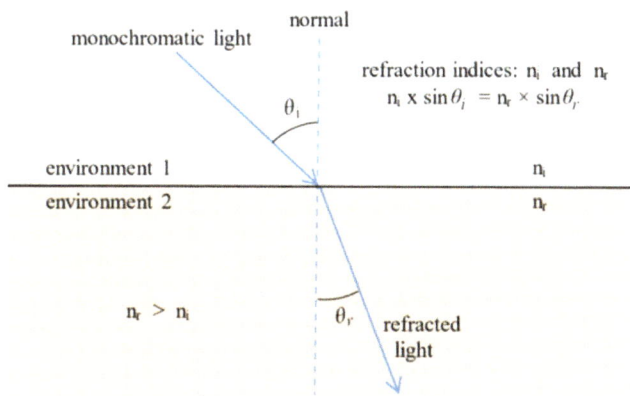

FIGURE 2.2   Refraction angle of a color.

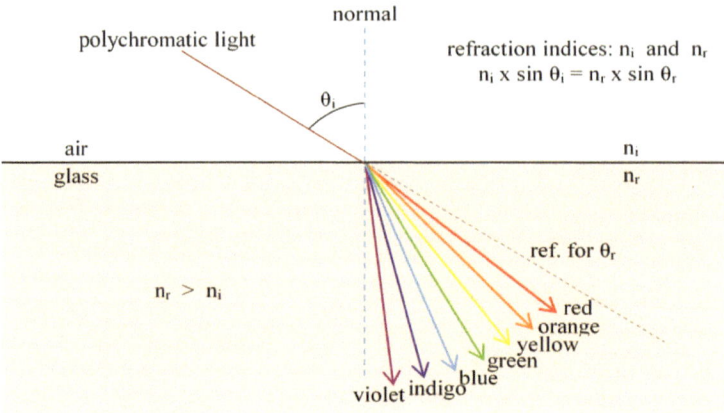

FIGURE 2.3 Refraction angles for visible colors.

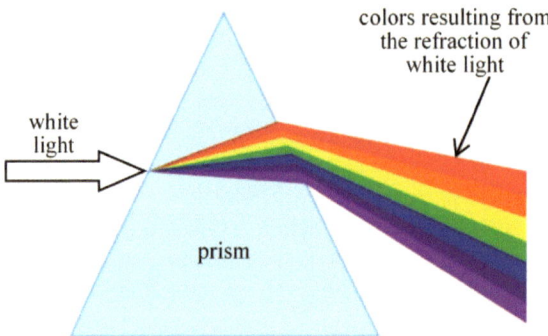

FIGURE 2.4 Angles of refraction of white light using an optical prism.

ability to detect even a single photon. Even though the retina is so sensitive to light, infrared and ultraviolet radiations go beyond the capacity of ordinary human eyes. In the case of infrared radiation, it passes through the eye and reaches the retina, starting a photochemical reaction, and only then will it be perceived by the brain. Laser beams are capable of emitting one or more extremely short wavelength bands, being a special source of infrared radiation.

IR is beyond ordinary human visual capabilities and is released by bodies that emit heat, such as the Sun and fire. In addition to these, humans are exposed to artificial sources, including heating devices and infrared lamps and saunas.

White light is made up of different light frequencies and the refractive index will depend on each of these frequencies, as shown in Figure 2.4 for a polychromatic light. This means that the angular deviation of the light is also different for each color.

According to what is presented in this section, it is to be expected that people do not in general perceive radiation identically. Some people have a brain capable of detecting a more or less wide range, slightly beyond or below colors and/or heat (thermal vision), light, sensations or sounds normally considered as such by ordinary mortals. A more limited example of this is color blind people who only have the ability to distinguish shades of brown, green or gray colors, varying with the amount of pigment in the focused objective. The tendency is for green to appear red, like sepia, as well as other distinct perceptions.

2.6 FREQUENCY SPECTRA AND THEIR EFFECTS ON LIVING BEINGS

As previously discussed, cone cells allow living beings to discriminate colors. An organism that has tetrachromatism is a tetrachromat because it has four types of cone cells for color discrimination, as it has a four-dimensional color space. In comparison, tetrachromats are endowed with a fourth type of cone, which is most sensitive in the yellow–green region of the visible spectrum. In other words, the visible spectrum of the tetrachromat can make a mixture of four primary colors, unlike trichromatic beings who use only three primary colors (RGB) and have lesser perception, as listed in Table 2.7. Therefore, tetrachromats are considered to be beings that see more color gradations. Many women are mainly able to see colors that are invisible

TABLE 2.7
Frequency spectra and some of the functions attributed to them

Range	Until	Service
20 Hz	20,000 Hz	Audible sounds
20 kHz	30 kHz	Ultrasound
530 kHz	1,600 kHz	AM radio stations with 10 kHz band
34.48 MHz	34.82 MHz	Taxi radio
38 MHz	40.6 MHZ	Biomedical telemetering
40.6 MHz	40.7 MHz	Telemeasurement of material characteristics
40.7 MHz	41.0 MHz	Biomedical telemetering
41.0 MHz	49.6 MHz	Various services
49.6 MHz	49.9 MHz	Cordless telephone
49.9 MHz	54 MHz	Various services
54 MHz	60 MHz	VHF television – Channel 2
60 MHz	66 MHz	VHF television – Channel 3
66 MHz	70 MHz	VHF television – Channel 4
70 MHz	72 MHz	Radio astronomy
72 MHz	73 MHz	Remote control
73 MHz	75.4 MHz	Aeronautical navigation radio
75.4 MHz	76 MHz	Remote control
76 MHz	82 MHz	VHF television – Channel 5
82 MHz	88 MHz	VHF television – Channel 6
88 MHz	108 MHz	FM radio – 99 channels in 200 kHz bands
88 MHz	108 MHz	Restricted range wireless microphone
108 MHz	117.975 MHz	Radio navigation for aeronautics
117.975 MHz	121.5 MHz	Mobile communication for aeronautics

(continued)

TABLE 2.7 (Continued)
Frequency spectra and some of the functions attributed to them

Range	Until	Service
121.5 MHz	121.5 MHz	Distress communication
121.5 MHz	136 MHz	Mobile communication for aeronautics
136 MHz	138 MHz	International meteorological satellites
138 MHz	143.6 MHz	Reserved for fixed and mobile communications
143.6 MHz	143.65 MHz	Space research
143.65 MHz	144 MHz	Amateur radio
146 MHz	148 MHz	Amateur radio
144 MHz	146 MHz	Amateur satellite radio
148 MHz	149.17 MHz	SESC – special supervision and control service
149.17 MHz	174 MHz	Various services
174 MHz	180 MHz	VHF television – Channel 7
180 MHz	186 MHz	VHF television – Channel 8
186 MHz	192 MHz	VHF television – Channel 9
192 MHz	198 MHz	VHF television – Channel 10
198 MHz	204 MHz	VHF television – Channel 11
204 MHz	210 MHz	VHF television – Channel 12
210 MHz	216 MHz	VHF Television – Channel 13
216 MHz	470 MHz	Various services
470 MHz	476 MHz	UHF television – Channel 14
476 MHz	482 MHz	UHF television – Channel 15
482 MHz	806 MHz	UHF television – Channels 16 to 69
806 MHz	824 MHz	Various services
824 MHz	834.4 MHz	A band cellular telephony
834.4 MHz	845 MHz	B band cellular telephony
845 MHz	869 MHz	Various services
869 MHz	880 MHz	A band cellular telephony
880 MHz	880.6 MHz	Other services
880.6 MHz	890 MHz	B band cellular telephony
890 MHz	891.5 MHz	A band cellular telephony
891.5 MHz	894 MHz	B band cellular telephony
894 MHz	896 MHz	Aeronautical cellular telephony
896 MHz	3,000 MHz	Other services
1 GHz	10 GHz	Microwave ovens
3 GHz	3.1 GHz	Radio navigation and radio location
3.7 GHz	4.2 GHz	C band satellite signal descent
5.925 GHz	6.425 GHz	C band satellite signal boost
6.425 GHz	7.125 GHz	Digital system
10.7 GHz	11.7 GHz	Digital radio
10.7 GHz	12.2 GHz	Ku band satellite signal descent
13.75 GHz	14.8 GHz	Ku band satellite signal boost
14.5 GHz	15.35 GHz	Digital radio

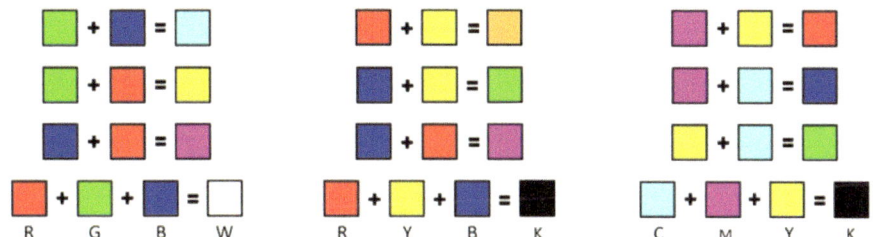

FIGURE 2.5 Color comparison: RGB, RYB and CMYK.

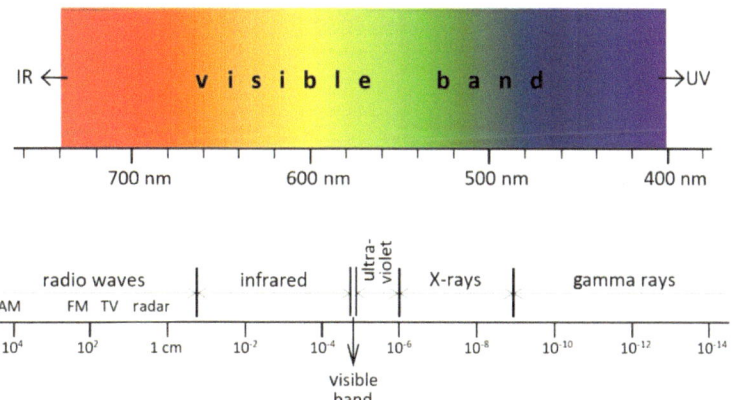

FIGURE 2.6 Wavelengths of the most known frequencies (in cm).

to other people thanks to a variation in a gene that influences the development of their retinas. They suffer or enjoy a condition known as tetrachromatism. The word tetrachromatism or tetrachromacy is a mixture of the words of Greek origin, 'tetra', which means 'four' and 'chromatic', which means 'color' [14,15]. There are also many cases of animals with senses that humans do not have.

Colors can be additive (light color) or subtractive (pigment color). In additive color (RGB), objects emit light (television, monitors, flashlights, etc.) with the addition of different wavelengths of the primary light colors, in this case red + blue (cobalt) + green forming the white color, see Figure 2.6. In subtractive color (cyan, magenta, yellow and black (CMYK)), a white surface is pigmented by mixing the secondary colors of light (primary colors in plastic arts): cyan + magenta + yellow. The letter K, in this case, is used to mean black because B cannot be used, as it already indicates blue. This is the color code used by electronic display systems.

Regarding the combination of colors, for CMYK it is noted that this is a 'subtractive system' where the colors are formed by absorption/subtraction of light on a white background in combination with the intensity of the added colored pigments. It is precisely the opposite of RGB, which emits light, so it has difficulty in faithfully reproducing all the colors on a screen monitor on paper. In CMYK there is already the white background color (white paper) and it is possible to obtain all the colors according to the addition of other colors, including black. An RGB combination cannot obtain this (Figure 2.5). The CMY system of printers and arts is used to obtain pigmented colors (non-light-emitting inks and objects). In this system, subtracting the three pigments (CMY) results in a very dark color tone, often confused with black. Even with the limitations of color selection, the RYB color model is still important to artists because it:

1) Explains the relationships between colors;
2) Helps to understand the changes in colors seen in nature;
3) Shows how to mix colors with predictable results;
4) Creates color palettes and color guides;
5) Organizes colors by name, codes and formulas;
6) Forms the basis of rules for the successful use of colors.

Currently, CMYK is considered the best subtractive model, capable of representing all colors perceptible by the human eye. RYB was historically used instead of the current CMYK because natural cyan and magenta pigments were very rare, which is why they were replaced, respectively, by blue and red, that is, it is necessary to remove colors to achieve the desired tone. O CMYK is capable of reasonably representing all colors perceivable by the human eye.

As James Clerk Maxwell (1831–1879) demonstrated in the 19th century, many colors can be reproduced with additive combinations of the three primary colors (RYB). So, for example, green and red produce the sensation of yellow. Blue and green produce the sensation of cyan (water green); red and blue produce the sensation of magenta (lilac or fuchsia). UV is the band of non-ionizing radiation that lies close to an ionizing radiation in the electromagnetic spectrum. Its main source is the Sun and its wavelengths naturally produced by the Sun can be classified in nanometers ($1 \text{ nm} = 10^{-9}.\text{m}$) as being: UVA (400–315 nm), UVB (315–280 nm) and UVC (280–100 nm).

2.7 INTERACTION BETWEEN FREQUENCY SPECTRA

All the senses of living beings (touch, hearing, vision, smell and taste) seem to depend on vibrations that are nothing more than the frequencies of oscillations received from the environment, as presented in Chapter 1. To summarize the frequency ranges, it can be said that those of touch are the lowest and depend on the differences in vibration between the touched parts, ranging from 0 to 200 cycles/second (Hz). The frequencies of human hearing or sounds can typically range from something like 15 Hz up to 25 kHz 25×10^3 Hz. The universally adopted standard ranges from 20 Hz to 20 kHz, which appears to be the normal range of frequencies perceivable by the vast majority of people.

As previously mentioned, the frequencies of vision are normally a combination of the frequencies of the basic colors: red, green and blue (RGB). These frequencies typically range from 405 THz (1 THz $=10^{12}.$Hz) (red color) up to 790 THz (violet color). The frequencies of smell and taste have to do with the particles present in the respective organs. Table 2.8 is a list of frequencies to which humans and animals are

TABLE 2.8
Low natural resonance frequencies of the body organs

Organs	Resonant low freq. (Hz)
Thoracic cavity	4–6
Abdominal	4–8
Shoulders	4–8
Lungs	4–8
Heart	5
Pelvic	6
Abdominal cavity	6–9
Whole body	7.5
Body torso	7–13
Spinal column	8
Head	8–40
Spine	10–12
Hands and arms	20–70
Chest wall	60
Ocular globe	60–90
Maxilla	100–200

TABLE 2.9
High natural resonance frequencies of the human body organs

Organs	Natural high freq. (MHz)
Liver	55–60
Disease start at	58
Pancreas	60–80
Human body	62–78
Heart	67–70
Normal brain	72
Brain	72–90

exposed, with effects still to be better studied. However, the widespread idea that there may be a so-called 'sixth sense' refers to an extrasensory perception, and is often based on a supposed spirituality. It is often said that women have a sixth sense; that is, they are much more acute than men.

There are several forms of manifestation of human existence through emissions from the human body into the environment. Here are included: image, heat, figure, movement, noise (voice, snoring, smile, shadow, smell, tapping ...); all of these are detectable manifestations. Obviously, any other living being will be subject to the frequencies of the environment in which it lives. In the human physique, each part of the body will always be subject to higher or lower frequencies [16–18]. In the case of the physical structure, the parts begin to resonate at a frequency that is not always the same as the others, as shown in Tables 2.8 and 2.9, causing different perceptions. These higher frequencies are generally related to human feelings such as fear, tiredness, sadness, anger, joy, surprise, shame, boldness, depression, predictions, culture and others [19–25].

It is interesting to say regarding the senses, that it is proven that people who suffer from a deficiency related to the sensory system end up developing other senses further. For example, a blind person develops their ability to hear or even feel books in Braille. Braille writing is done from right to left with dots created by punches. To read, the blind person turns the page over and uses the relief formed on the back. The question then arises: are there other frequencies that can be felt unconsciously by humans and animals, that is, are there other unknown sensors within the human body [14,15]?

When it comes to measuring vibrations with higher frequencies, it is common to use the unit known as the electron volt (eV). An electron volt is the amount of kinetic energy gained by a single electron in a vacuum when accelerated by a potential difference of 1 volt. X-rays are photons with energy between 100 eV and 100 keV, while gamma rays are above 100 keV. Recent evidence indicates that both radiations can be generated by electrons accelerated, respectively, by the electric field existing within the channel of a lightning bolt (X-ray) and a cloud (gamma ray). Figure 2.6 shows most of the best-known frequency bands in the Universe.

2.8   KIRLIAN EFFECT

In 1939, Semyon Davidovich Kirlian accidentally discovered that if an object is placed on a photographic plate subjected to a large potential difference, high frequency and low electrical current, it leaves an image printed on the plate. This phenomenon became known as the Kirlian effect [26,27]. In reality, Kirlian photography records corona discharge by a more or less heated body commonly involving any conductive object that is placed in a dielectric enclosure at atmospheric pressure within a high-voltage/high-frequency electric field. Included among the conductive objects is the structure of the human body, which for the most part is made up of salt and water and, therefore, is a good conductor of electricity.

An example of the use of X-rays is the fact that the bones can absorb them, and therefore it is possible to produce images of the inside of the human body (see Figure 2.7). Gamma radiation is harmful to living tissues because it has ionizing power, but medicine uses it for radiotherapy, directing it with reasonable energy control to destroy malignant tumors in patients, preserving healthy tissues in the rest of the body [23].

Old school Kirlian photography required contact printing by placing a sheet of photographic paper on a metal plate. Then, the subject of the desired photo was placed on top of the paper (see Figure 2.8). A high voltage frequency passed through the plate with the electrodes attached to the metal plate. The process created what is known today as a corona discharge that occurs between the subject and the metal plate. Once the photographic paper is developed, the resulting image shows the

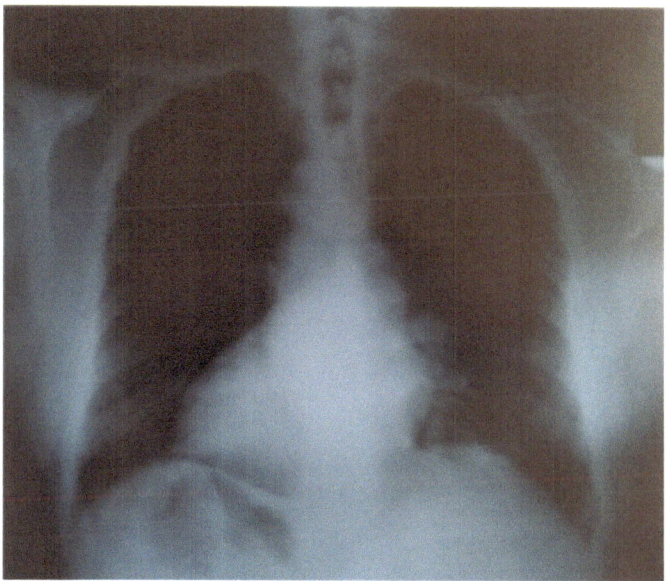

FIGURE 2.7   Image of the inside of the human body.

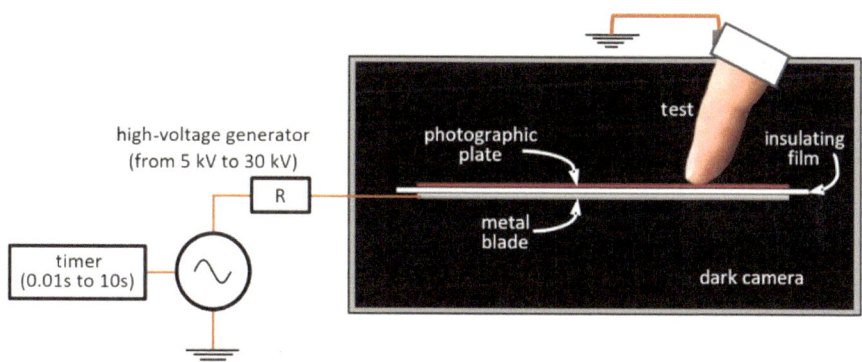

FIGURE 2.8   Basic scheme for Kirlian photography.

subject with a glowing silhouette or aura around its edges. Some examples of Kirlian photographs are displayed in [24].

The general appearance of Kirlian discharge depends on many factors and can change continuously on a small scale. Reproducibility is therefore poor. There is no satisfactory proof of the claims that the crowns around the fingertips reflect the physical or mental health of the subject or that they may reveal psychic influences.

Humidity is the main determinant of the shape and color of Kirlian photographs of human beings and is one of the most 'mysterious' photographic genres, surrounded by myths and a good deal of science. Such myths are commonly associated with photographs that intended to capture the aura of all living beings, including humans.

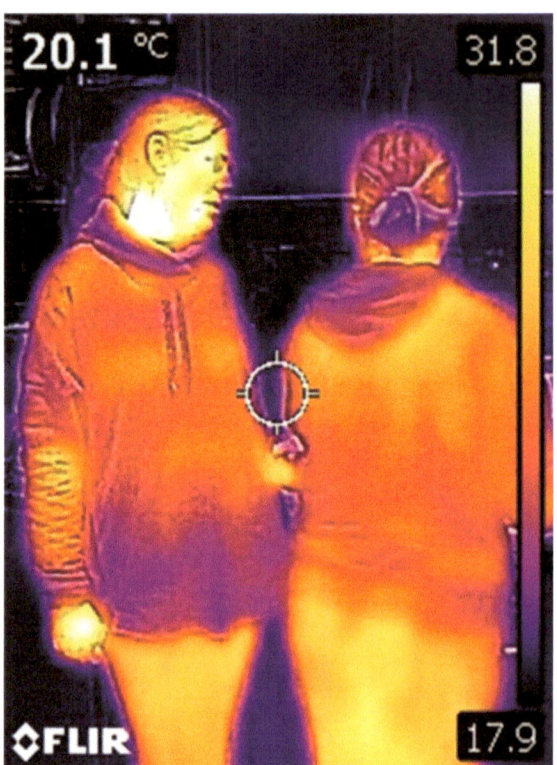

FIGURE 2.9   IR photography of family members (Courtesy of Eng. Jairo Pires – 29 Jun 2022).

As is to be expected, some people from childhood notice more or less broad intensities of colors, sounds or heat around things, animals or other people that may reflect physical ailments in what they see. People with headaches, fever, stomach upset or any other organic disorders change their spectrum of colors and heat around the body, better known as aura, which can indicate what is happening to them. Therefore, as some children grow up with this unexplainable sensitivity, they associate the heat and colors of other people's auras with human indispositions, unlike ordinary mortals who do not have this sensitivity. Other people are surprised by this ability, which may usually be recognized in some religions as mediumship or clairvoyance.

It should be noted here, for example, that the interpretation of changes in the heat spectrum is also widely used by maintenance engineering teams for electrical power transmission lines who need to detect the aura around electrical wires to locate heat generated by possible connection defects. Similarly to people capable of perceiving the aura, these teams use infrared radiation detectors to locate any heat that may be caused when possible defects occur in electrical wiring connections. Furthermore, several other devices also work based on IR radiation, such as remote controls. These controls use a light-emitting diode (LED) that emits IR radiation, which is then detected by a sensor on the electronic device, such as automatic doors, televisions, barcode readers and computer mice [2,3].

2.9   CORONA EFFECT

The corona effect is the result of intense and high contact of an electric field with particles of air, humidity or dust, causing them to emit light whenever they are ionized. This is the case when ionization of the air around high-voltage transmission lines causes the conductors to glow, producing a hissing noise known as the corona effect. These discharges or corona effects happens when an outdoor object is filled with enough electrical charge that the electrical charge leaks from the object into the surrounding air, becoming electrified and producing a characteristic glow.

The corona effect occurs continuously, increasing with line voltage and frequency. The voltage necessary for the beginning of its manifestation decreases as the frequency increases, and can occur at voltages as low as 300 V. The corona effect also appears in living beings (plants, animals), and even metals, due to the ionization of the air around them caused by high-frequency, high-voltage and low-amperage currents, plus sweating and gases (body vapors) associated with temperature [23]. In Figure 2.9 one can clearly perceive the visual colors purple, red, yellow and white as being the notable differences in a person's aura.

Nowadays, there are commercial cameras that make it possible to record manifestations of the corona around objects [27].

REFERENCES

[1]   E.R. Kandel, J.H. Schwartz, and T.M. Jessell, Principles of Neural Science, 4th ed. McGraw-Hill, 2000, p. 433.

[2]   E. Duane Haines, Neurociência Fundamental: com aplicações básicas e clínicas. 3rd ed. Elsevier, ISBN: 9788535219777, 2006.

[3]   K. Ochi, K. Nakakura-Ohshima, S.H. Youn, S. Wakisaka, and T. Maeda, The Ruffini ending as the primary mechanoreceptor in the periodontal ligament: Its morphology, cytochemical features, regeneration, and development, Critical Reviews in Oral Biology & Medicine, 10(3), 307–327. PMID: 10759411, Doi: https://doi.org/10.1177/10454411990100030401, 1999.

[4]   M.L. Machado Duarte and M. de Brito Pereira, Vision influence on whole-body human vibration comfort levels, Shock and Vibration (IOS Press), 13, 367–377, ISSN 1070-9622/06/$17.00, 2006.

[5]   H.E. Von Gierke and A.J. Brammer, Chapter 44: Effects of shock and vibration on humans, in Cyril M. Harris (ed.), Shock and Vibration Handbook, 4th ed. McGraw-Hill, 1998.

[6]   M.R. Misael, Human Comfort to Vibration Levels (in Portuguese), M.Sc. Dissertation, Mechanical Engineering Department, Federal University of Minas Gerais, Belo Horizonte/MG, Brazil, 2001.

[7]   P. Coffey, The Science of Logic: An Inquiry into the Principles of Accurate Thought. Longmans, 1912.

[8]   S. Isacoff, Temperament: How Music Became a Battleground for the Great Minds of Western Civilization. Knopf Doubleday Publishing Group. 16 Jan 2009, pp. 12–13. ISBN 978-0-307-56051-3. Retrieved 18 Mar 2014.

[9]   D.E. Haines, Neurociência Fundamental: com aplicações básicas e clínicas. 3rd ed. Elsevier, 2006.

[10] J. Bolton and C. Colton, The Ultraviolet Disinfection Handbook. American Water Works Association. ISBN 978-1-58321-584-5, 2008.

[11] P.E. Hockberger, A History of Ultraviolet Photobiology for Humans, Animals and Microorganisms, Photochemistry and Photobiology, 76(6), 561–569, https://doi.org/10.1562/0031-8655(2002)0760561AHOUPF2.0.CO2, 2002.

[12] D.M. Hunt, L.S. Carvalho, J.A. Cowing, and W.L. Davies, Evolution and spectral tuning of visual pigments in birds and mammals, Philosophical Transactions of the Royal Society B: Biological Sciences, 364(1531), 2941–2955. https://doi.org/10.1098/rstb.2009.0044, 2009.

[13] P.E. Hockberger, A history of ultraviolet photobiology for humans, animals and microorganisms, Photochemistry and Photobiology, 76(6), 561–569. https://doi.org/10.1562/0031-8655(2002)0760561AHOUPF2.0.CO2, www.thoughtco.com/definit ion-of-ultraviolet-radiation-604675, 2002.

[14] BioExplorer.net. Top 11 Animals with Super Sensors, Bio Explorer, www.bioexplo rer.net/animals-with-best-sensors.html/, 14 Aug 2023.

[15] C.J. Ferraris, Jr., C.D. de Santana, and R.P. Vari, Checklist of Gymnotiformes (Osteichthyes: Ostariophysi) and catalogue of primary types, Neotropical Ichthyology, 15(1). https://doi.org/10.1590/1982-0224-20160067, 2017.

[16] C.W. Ren, B. Peng, J. Shen, Y. Li, and Y. Yu, Study on vibration characteristics and human riding comfort of a special equipment cab, Hindawi Journal of Sensors, Article ID 7140610, 8 pages, https://doi.org/10.1155/2018/7140610, 07 Feb 2018.

[17] M.L. Machado Duarte and M. de Brito Pereira, Vision influence on whole-body human vibration comfort levels, Shock and Vibration (IOS Press), 13, 367–377, 2006.

[18] A. Sharma and A.K. Maurya, International Journal of Electrical, Electronics and Data Communication, Health and Bass, www.healthandbass.com/research. 20 Aug 2021.

[19] A. Sharma and A.K. Maurya, Aggregate frequencies of body organs, International Journal of Electrical, International Journal of Electrical, Electronics and Data Communication, Health and Bass, 5(11), ISSN: 2320–2084, http://iraj.in, Nov 2017.

[20] N. Daimiwal, M. Sundhararajan, and R. Shriram. Respiratory rate, heart rate and continuous measurement of BP using PPG. In 2014 International Conference on Communications and Signal Processing (ICCSP). IEEE, 2014, pp. 999–1002.

[21] T. Sato, and Y. Watanabe, High sensitivity estimation of red blood cell aggregation with ultrasonic peak frequency. In 2013 IEEE International Ultrasonics Symposium (IUS). IEEE, 2013, pp. 868–871.

[22] G. Çetin, G. Akkulak, and S. Özdemir, Locate the internal organs in the human body: A survey in Turkey, Procedia-Social and Behavioral Sciences, 116, 2819–2824, 2014.

[23] R. Helerbrock, Espectro Eletromagnético, Brasil Escola. Disponível em: https://brasi lescola.uol.com.br/fisica/espectro-eletromagnetico.htm. Accessed 12 May 2023.

[24] K.L. Johnson, The Living Aura: Radiation Field Photography and the Kirlian Effect. Toronto Public Library, 1975, 178 p.

[25] R. Gerber, Vibrational Medicine, New Choices for Healing Ourselves. Bear & Company, ISBN: 13-978-0939680467, 01 Jan 1988.

[26] J. Russell and R. Cohn, Semyon Davidovich Kirlian. Amazon.com, ISBN-10: 5512796461, ISBN-13: 978-5512796467, 11 Jan 2013.

[27] J. Opalinski, Kirlian-type images and the transport of thin-film materials in high-voltage corona discharges, Journal of Applied Physics, 50(1), 498–504. https://iee explore.ieee.org/xpl/articleDetails.jsp?arnumber=5105453, Jan 1979.

3 Electromagnetic Waves and Living Beings

3.1 INTRODUCTION

The objective of this chapter is to show how electromagnetic emissions can affect living beings and, in particular, humans. It is shown that in addition to the natural frequencies discussed in Chapters 1 and 2, living beings are exposed to intense artificial electromagnetic frequencies used by humans, mainly for communications and medicine. With these frequencies, nowadays, living beings of all species are constantly exposed to intense densities of electromagnetic radiation regardless of their willingness. Such frequencies are distributed intensively all over the world. The sources of these radiations are mainly used in cell phones, television systems, radio stations, signal detectors in public environments and medical equipment, among many others.

3.2 ELECTROMAGNETIC WAVES

The expression 'electromagnetic radiation' (emission) refers to any energy emitted as wave forms. Therefore, radiation (emission) is the number of photons emitted by a single source, while irradiation (reception) refers to the radiation that falls on a surface exposed to it, affecting all living beings in its range. For example, solar radiation irradiates planet Earth, providing it with energy, mainly in the forms of heat and visible light.

Vertical antennas with voltage applied between their bases and the ground are widely used to distribute frequencies in megacycles per second. The radiation received at any distant point, such as P shown in Figure 3.1, will comprise an unchanged direct component, and a reflected component. If the ground reflections were perfect, it could be considered a ray reflecting off the ground and coming from an image antenna. The problem of radiation from an antenna and its image is simpler than the problem of an antenna as a reflecting plane [1].

Reflection from the Earth's surface is not perfect, but it can be quite good for typical practical values of frequency and conductivity. If the Earth's surface were a perfect conductor, the reflection coefficient would be −1 for all angles of incidence, because the sum of the incident and reflected tangential components of the electric field in this case should always be zero. In other words, the horizontal component of the electric field would be completely inverted by reflection (see Figure 3.1). The

DOI: 10.1201/9781003604037-3

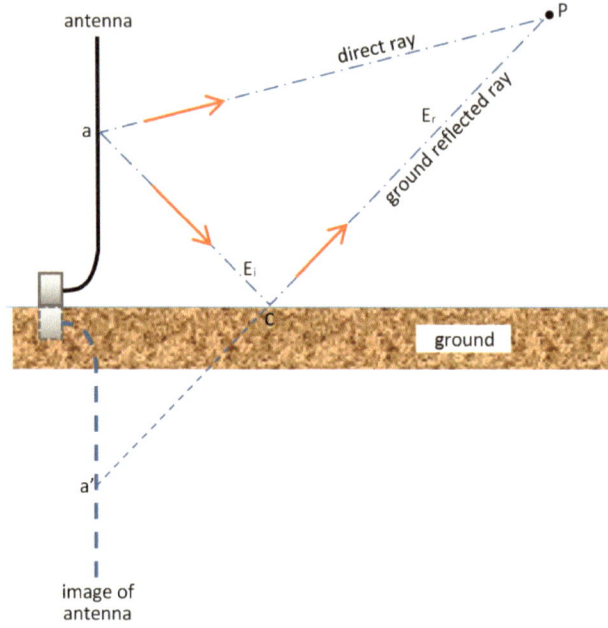

FIGURE 3.1   Radiation directions of grounded antennas.

vertical component would not be inverted, since, as can be seen in the diagram, an unchanged vertical component E_i is necessary to follow an inverted horizontal component E_r after reflection.

If, on the other hand, the Earth were a perfect dielectric material, the reflected wave E_r would be smaller than the incident wave, E_i, with the magnitude of the reflection coefficient being very dependent on the angle of incidence and soil conductivity. In fact, at a certain angle, known as the Brewster angle, all the energy that reached the surface would enter the dielectric and the reflection would be zero. The Brewster angle is the angle of incidence that is formed when an incident wave with a certain polarization is transmitted perfectly by a flat surface of a refracting system that presents two transparent and homogeneous media separated by a surface, known as a diopter. That is, in Brewster's law, $n_1 \cdot \sin\theta_i = n_2 \cdot \sin\theta_r$, where n_1 e n_2 are the refractive indices of media 1 and 2 and θ_i and θ_r are the angles of incidence and refraction, respectively. When an unpolarized incident wave hits the Brewster angle, the reflected ray is then perfectly polarized, as shown in Figure 3.1. For this reason, the Brewster angle is also known as the polarization angle. The existence or not of phase inversion of the reflected ray depends on whether the angle of incidence is greater or less than the Brewster angle [1].

3.3   ELECTROMAGNETIC INDUCTION

Electromagnetic or magnetic induction is manifested by the production of an electromotive force (emf) in an electrical conductor caused by a variable magnetic field.

This emf can produce an electrical current of greater or lesser intensity. Figure 3.2 is a representation of an alternating electrical current flowing in the coil conductor shown on the left. This current produces a variable magnetic field φ, which electro-magnetically induces an electric current i, which flows in the conducting loop on the right. The coil on the left side is responsible for the electric field that surrounds the coil (solenoid) on the right side. Analogous to this phenomenon, the human body is subject to this type of induced current because it is made up of water and salt; that is, it is an electrical conductor.

A parallel can be made here between the cell phone's transmitting antenna represented synthetically by the coil (current i_1) on the left of Figure 3.2 and the conductor coil on the right consisting of water and salt (current i_2). Currents will circulate in it that can produce heating or burning that will depend on the intensity of the induced currents, which, in turn, depend on the intensity and frequency of the magnetic field that created them.

3.4 FARADAY'S INDUCTION LAW AND LENZ'S LAW

To give a scientific basis to the phenomenon of magnetic induction, imagine that a constant electric current circulates through the longitudinal cross-section of the solenoid shown in Figure 3.2. The lines shown in this figure will be those of the magnetic field, with their directions represented by the blue arrows. The magnetic flux corresponds to the 'density of the field lines' and is therefore denser in the middle of the solenoid and weaker outside it. Mathematically, Faraday's law of induction makes

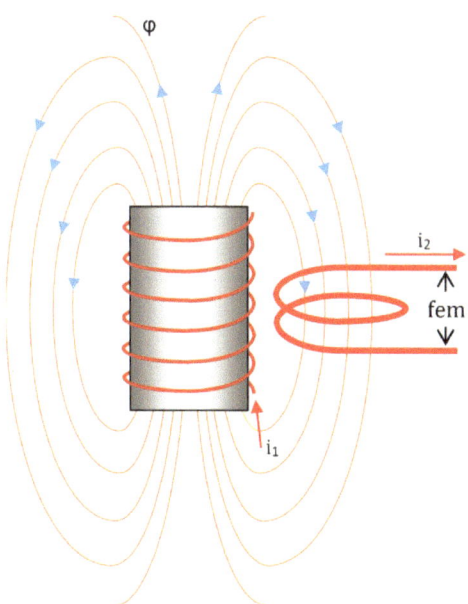

FIGURE 3.2   Distribution of magnetic radiation in the space.

use of the magnetic flux φ in the region of the space delimited by the conductor turns. The magnetic flux will then be defined by a surface integral given by [1,2]:

$$\varphi = \int B dA \qquad (3.1)$$

where dA is an element of the surface delimited by the conductor turns and B is the magnetic field.

The scalar product BdA corresponds to an infinitesimal amount of magnetic flux. In more visual terms, the magnetic flux through the conducting loop is proportional to the number of magnetic field lines that pass through this loop. When the flux through the surface varies, Faraday's law of induction quantifies an emf on the conducting loop. The most widespread version of this law states that the emf induced in any closed circuit will be proportional to the rate of variation (frequency) of the magnetic flux enclosed by the circuit. This includes the blood of humans, which is mainly made up of the electrically conducting combination of water and salt:

$$e = -\frac{d\varphi}{dt} \qquad (3.2)$$

where e is the emf and φ is the magnetic flux.

For the direction of the emf e, Lenz's law states that an induced current i will flow in the opposite direction to the change that produced it [1,2]. This is represented by the negative sign used in equation 3.3. Increasing the number of turns N, there will be an increase in the emf generated, in a composition of N identical turns with a total ohmic electrical resistance, R, each passing through the same magnetic flux. The resulting emf will then be N times that of a single driver.

$$e = -N\frac{d\varphi}{dt} = Ri \qquad (3.3)$$

The generation of an emf by a variation in the magnetic flux on the surface of a conducting loop i can be achieved in several ways, whether by varying the magnetic field, deforming the loop, changing the orientation of the loop relative to the direction of the magnetic field, or some combination of these. The relationship between the emf e in the conducting loop surrounding a surface A and the electric field E in the conductor will be given by the Maxwell–Faraday equation as:

$$d(e) = Ed\ell \qquad (3.4)$$

where dℓ is a surface boundary element A, combining this with the definition of the flow:

$$\varphi = \int B dA \qquad (3.5)$$

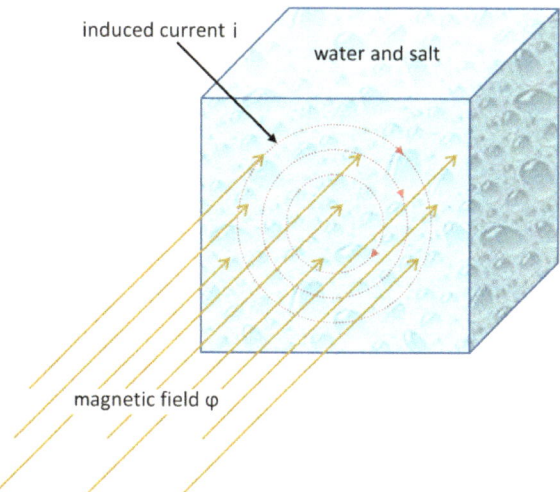

FIGURE 3.3 Current induced in salt water (blood).

As a result, one can write the integral form of the Maxwell–Faraday equation, which plays a fundamental role in the theory of classical electromagnetism as:

$$\int Ed\ell = -\frac{d}{dt}\left(\int BdA\right)$$
(3.6)

With regard to electrical conductivity, it must be considered that the human body is made up of an average of 60% water and approximately 0.15% salt, which is the electrolyte that conducts the electricity necessary for cells, muscles and the nervous system to function. The distribution of salt in the body varies depending on the tissue. Adipose tissue contains virtually no water, but skeletal muscles are made up of something like 73% water. Blood plasma is made up of more than 90% water. Sweat constantly alters these proportions, but is responsible for maintaining a constant body temperature of approximately 37 °C. In short, the human body is a good conductor of electricity and, therefore, subject to electromagnetic induction of electrical currents arising from the distribution of radiation from antennas in the environment, as illustrated in Figure 3.3.

3.5 DISTRIBUTION OF RADIATION WITH DIRECTED ANTENNAS

Figures 3.4 and 3.5 are graphic representations of how electromagnetic waves from directional antennas are distributed in the environment. The distributions of these waves are illustrated as seen from above and from the side. This type of directional wave is mainly used to carry electromagnetic signals directly from one antenna to another, concentrating on well-defined points.

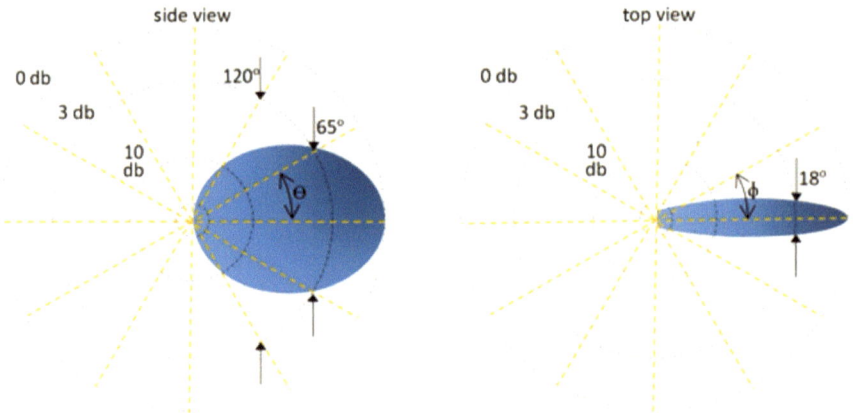

FIGURE 3.4   Directive antenna with its radiation diagrams from side and top views.

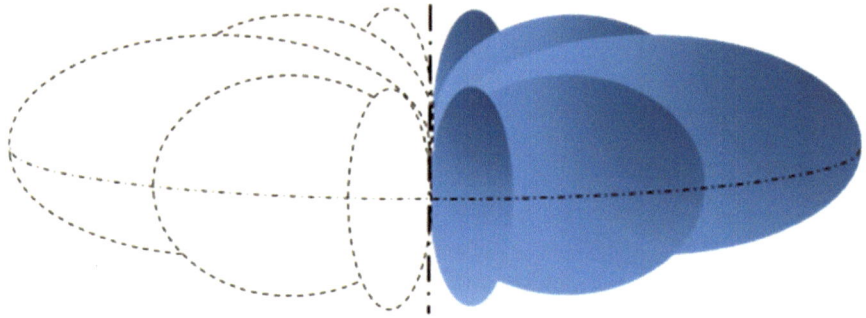

FIGURE 3.5   Distribution of the radiation directions.

The power density P and the intensity of the electric field E around the antenna arriving at the reception point can be defined as [1]:

$$P(r,\theta,\varnothing) = \frac{1}{r^2}F(\theta,\varnothing)$$

(3.7)

$$E(r,\theta,\varnothing) = \frac{1}{r}G(\theta,\varnothing)$$

(3.8)

where the variable r represents the distance from the base of the antenna to the point of measurement of radiation intensity, θ is the direction of horizontal propagation and ∅ is the angle of vertical elevation [1,2].

Directional antennas in a given environment face a single selected direction. The signs P and E of these antennas are strong and therefore their use is common to connect the internet as they concentrate all the signal power in a specific direction, thus guaranteeing much better quality than would be the case with a more restricted coverage. Despite this, they also have their limitations, as these sources require a good position if a signal is needed that will meet Wi-Fi coverage, for example. Otherwise, they will not work correctly. This is one of the reasons why many people complain that in their homes the signal is great in a certain area and not in others. For small places that have just one room, this type of connection is very good, but if there are many partitions and rooms, the signal may become weak or unstable.

One of the biggest disadvantages of the directional antenna is that it cannot serve larger locations or those with many rooms or walls. A router with a directional antenna with coverage of 200 m² can serve this location perfectly, but only in a specific direction. This is due to the fact that these directional antennas, including parabolic and Yagi antennas, have a radiation pattern focused in a specific direction. There are also high-directional antennas that are designed to focus energy in a very specific direction, being useful only for long-distance communications. Table 3.1 lists some of

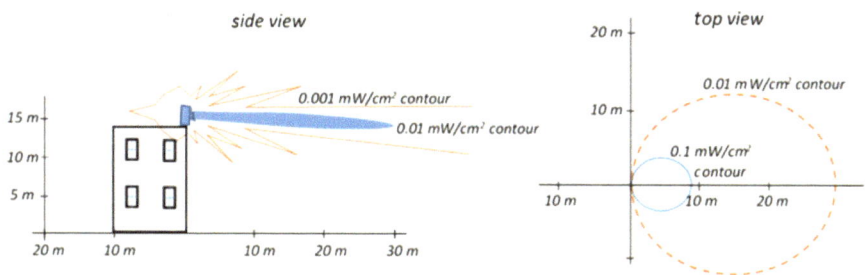

FIGURE 3.6   High gain directional antennas (1 μW/cm² = 0.01 W/m²).

TABLE 3.1
Standard power densities for data transmission

Standard	Power density (W/m²)	
	900 MHz	1800 MHz
ICNIRP	4.5	9.0
ANSI/IEEE	6.0	12.0
ANATEL	4.5	9.0
Italy	0.025	0.025
Switzerland	0.042	0.095
Salzburg	0.1 (0.001)	0.1 (0.001)
Campinas-SP-Br	1.0	1.0
Porto Alegre-RS-Br	4.5	9.0
Santa Maria-RS-Br	1.0	1.0

Sources: Maximum power density according to standards (f/200 W/m²); [3].

TABLE 3.2
Standards for minimum distance for data transmission systems

	Minimum distance (m)	
Standard	900 MHz	1800 MHz
ICNIRP	21.28	15.05
ANSI/IEEE	18.43	13.03
ANATEL	21.28	15.05
Italy	285.46	285.46
Switzerland	220.24	146.44
Salzburg	1,427.30	1,427.30
Campinas-SP-Br	45.14	45.14
Porto Alegre-RS-Br	220.24	146.44
Santa Maria-RS-Br	21.28	15.05

Sources: Minimum distance when the point of interest is in the main lobe (ANATEL: for 55 dBm and antenna gain of 15, dBi = 70 dB); [3].

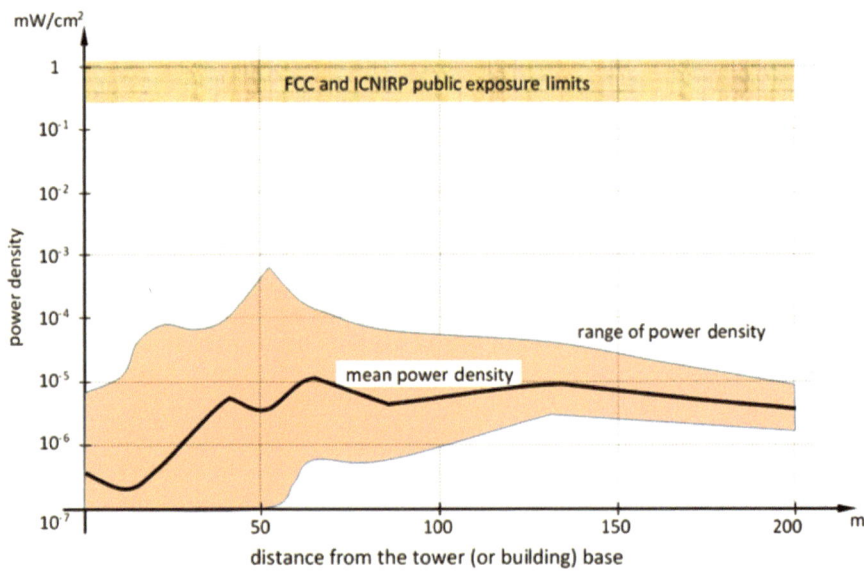

FIGURE 3.7 Reduction in power density with distance from the antenna.

the most important standards for using the 900 MHz and 1800 MHz frequencies that define the power densities allowed by operators [3–5].

Figure 3.6 is an example of how the intensities of electromagnetic waves are distributed, seen from above and from the side, as they are generated in directional antennas on top of buildings. It can be seen in this figure that as the propagation of the wave moves away from the direction required for its transmission, it decreases sharply in intensity [6,7].

According to the FCC and ICNIRP public exposure limits, a minimum acceptable distance from the antenna tower is necessary for the safety of living beings, achieving an appreciable reduction in radiation intensity (see Table 3.2). Figure 3.7 illustrates how this reduction in power density can occur depending on the distance from the antenna [9,10].

3.6   DISTRIBUTION OF RADIATION BY LOW-GAIN OMNIDIRECTIONAL ANTENNAS

Omnidirectional antennas have a relatively uniform gain in all directions in the horizontal plane and, therefore, are widely used in Wi-Fi routers. Hence, and unlike the directional antenna, the omnidirectional antenna can transmit signal in all directions around it, as shown in Figure 3.8. This is why it is recommended for very large places or those with many barriers, such as trees, walls and abundant furniture. Even though this antenna can transmit signals in all directions, the intensity with which it arrives is not as high as that of directional antennas because they spread in all directions, both horizontal and vertical. Therefore, in larger cities it is necessary to install several of these antennas.

The main feature of omnidirectional antennas is that they can transmit the signal 360°, but they do not concentrate much with the Wi-Fi signal. For wider places, even if the concentration of the Wi-Fi signal is not as strong, it is the best option, because when it is installed in the center of the room, everything else can have a good connection.

Unlike the radiation from the directional antenna, the electromagnetic wave from the low-gain omnidirectional antenna propagates in all directions perpendicular to its axis. This type of antenna is easy to use as it does not require direction, greatly simplifying its installation. It is used in both base stations and access cards. Figure 3.8 illustrates how this distribution of the electromagnetic field around this antenna happens.

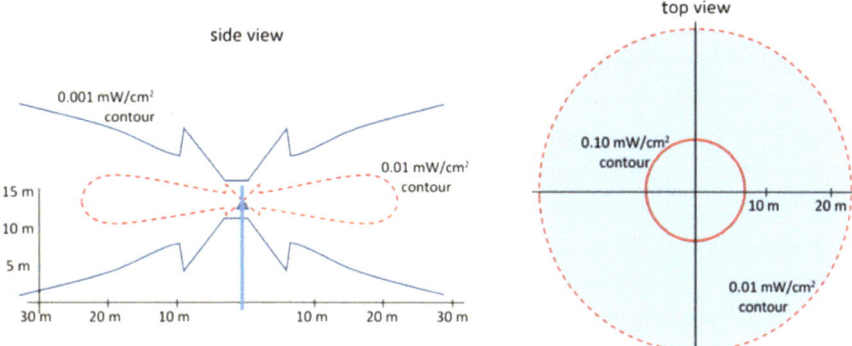

FIGURE 3.8   Low-gain omnidirectional antennas.

Because it is multidirectional, the omnidirectional antenna is easy to install and the resident or a technician, even with little experience, can do this, while it can be mounted on the roof of the residence upside down in such a way as to facilitate the distribution of the signal throughout the area.

A crucial factor for directional antennas is signal gain, that is, their ability to direct or focus transmitted energy in a specific direction, compared to an omnidirectional antenna that radiates energy evenly in all directions. Signal gain determines the distance and quality of communication that an antenna can transmit or receive signals over longer distances by focusing in a specific direction. On the other hand, a low-gain antenna may be preferable in applications where coverage in all directions is required.

3.7 FORMULAS FOR CALCULATING ANTENNA POWER GAIN

The power gain $G(dBi)$ of an antenna can be calculated by:

$$G(dBi) = 10 \cdot \log(G) \tag{3.9}$$

The dB unit stands for decibels, and is often used to express in milliwatts (mW) the strength of signals in wires and cables. In communications, this unit measures the strength of the signal reaching the destination. In particular, dBm means milliwatt decibels and dBi means signal gain in decibels in wires and cables related to an isotropic gain, as shown in Table 3.3.

When the values to express an absolute power range from minimum levels to extremely high values, decibel is a unit of measurement used mainly in telecommunications through a logarithmic relationship necessary to reduce the size of the representation. In this case, this unit is defined as the power level given in decibels relative to a standard reference level of 1 mW [1].

TABLE 3.3
Types of signal gains measured in dB

dB	dBm	dBi
dB stands for decibels	dBm stands for decibel milliwatts	dBi stands for decibel relative to an isotropic gain
Measure of loudness	Unit used to express decibels in milliwatts, often used to measure signal strength in wires and cables	It draws a comparison between a real antenna and a hypothetical isotropic antenna
dB gives the ratio	dBm gives absolute power	dBi to measure the strength of a hypothetical antenna
$L_p = 10 \cdot \log_{10}(P/P_o) dB$	$S(dBm) = 10 \cdot \log_{10} P$	$G(dBi) = 10 \cdot \log_{10} G$
dB is used to measure source intensity gain	Measures small values and is used in wires	dBi used to measure antenna performance

As discussed above, decibel is a very useful mathematical representation to also measure the difference between sound levels, and can be defined as $V - V_0 = 10 \cdot \log_{10}\left(V/V_0\right)$ to relate a level V with the level taken as reference, V_0. However, this unit can be used for any other type of signal reference gain. For example, the variable V_0 could be the sound level taken as a reference for a measurement. In this case, dBm would mean decibels milliwatts to measure the strength of V in relation to the reference. A higher dBi means a more direct gain from an antenna, which means the signal strength is greater, but the direction will be stricter, as illustrated in Figures 3.4 and 3.5. Therefore, the signal will not spread widely in a given direction.

The ideal gain of an antenna depends on the application. Inside a home it needs uniform coverage and in this case the omnidirectional antenna is recommended. Conversely, for long-distance point-to-point communications, a directional antenna with high gain is the best choice. Therefore, antenna gain is essential in determining wireless communication performance.

3.8   REGULATIONS FOR RADIATION

The International Commission on Non-Ionizing Radiation Protection (ICNIRP) presents the framework, activities and general approaches for providing guidance on protection against non-ionizing radiation (NIR). An independent description of ICNIRP could be the Best Practice Guidance System on the Protection of People and the Environment from Exposure to Non-Ionizing Radiation [6–9]. This Declaration highlights the independence of ICNIRP and presents the principle and requirements for the absence of commercial or other interests. ICNIRP's funding arrangements and collaboration with other advisory bodies and radiation protection authorities are also described. The statement also presents the types of guidance documents produced by ICNIRP and the general approach to evaluating scientific evidence.

Table 3.1 brings together the main regulatory standards, listing the minimum distances from the antenna base when the point of interest is in the main antenna lobe to be subject to electromagnetic effects [6–9]. When a high-frequency alternating current is applied to the human body, it acts like an antenna in transmission mode, radiating part of the energy and dissipating the rest as heat. Within the resonance region of the human body (< 200 MHz), one of the areas of study that involves the application of RF current to the human body is human body communication (HBC), which is a relatively new wireless communication technique using the human body as a communication channel to connect wearable electronic devices. Specifically, there are studies on two types of HBC, galvanic and capacitive coupled. In these studies it is shown that the magnitude of interference currents induced within the human body due to external electromagnetic fields is also maximum at the resonance frequency of the human body [11–14].

Table 3.4 lists the main effects of electromagnetic radiation from communication systems on living beings [15]. Although there are a large number of studies on the interaction between electromagnetic fields and the human body, little is known about the nature of the human body's antenna, much less about its application as an antenna. In this text, the human body is characterized theoretically and experimentally as a monopole antenna, in the frequency range of 10–110 MHz, to investigate

TABLE 3.4
Maximum power density according to standards (f/200 W/m²) (f = frequency in MHz)

Function investigated	Intensity rad. (µW/cm²)	Change types	Researchers
Body weight	150	Weight delay (continued experiment)	V.V. Markov
Blood pressure	150	Biphasic course with significant hypotension (continued experiment)	V.V. Markov
Reproductive function	150	Decrease in fertility and litter size, increase in defects in the number of offspring and embryonic mortality, etc. (continued experiment)	A.N. Bereznitskaya et al.
Central nervous system	10–20 and higher	1) Changes in EEG with predominant synchronization (immediate experiment)	Z. V. Gvozdikcova et al.
	150	2) Bivariant changes with activation predominance (immediate experiment)	
	150	3) Bivariant changes in basal sub-cortical structures (continued experiment)	
Electromyography	150	Aumento da atividade elétrica da unidade ativa	V.V. Markov
Hypothalamus–adrenal cortex system	150	1) Changes in the weight of the pituitary and adrenal endocrine glands 2) Changes in the neurosecretory function of the hypothalamus 3) Tendency of increased norepinephrine levels in the adrenal glands	Z. V. Gvozdikcova et al.
Metabolism	150	Changes in water and electrolyte metabolism (Na, K, water and total nitrogen excretion)	Z. V. Gvozdikcova et al.
Immunology	150	Inhibition of neutrophil phagocytic activity	A.P. Vokova & V.V. Markov

Source: [15].

its effectiveness as an antenna. The reflection coefficient is measured using a human subject as a monopole antenna. Theoretically, it is predicted that this human body can function as an efficient antenna with a maximum radiation efficiency reaching 70%, which is supported by many measurement results found in the literature [15,16]. This human body is found to resonate from 40–60 MHz depending on body posture when it is powered by a 50 Ω impedance system at the base of the foot. Therefore, a minimum reflection coefficient of −12 dB measured shows that the human body can potentially be used as an antenna.

On the other hand, a study stimulated by the German Federal Agency for Radiation Protection examined the risk of developing cancer in people who live within 400 m of cell phone antennas and in people who live further away. A set of 1000 patients were evaluated between 1994 and 2004 who lived up to 400 m from an antenna that has been operating in Naila since 1993. It was concluded that these people had three times the incidence of cancer than those who lived further away from the antenna [17].

The American National Standards Institute (ANSI) is not exactly a standard, but a private non-profit American organization, founded in 1918, dedicated to supporting US voluntary standards, conformity assessment system and strengthening their impact, whether in the USA or around the world. Similarly, the National Telecommunications Agency (in Portuguese: Agência Nacional de Telecomunicações; ANATEL) is a special agency in Brazil created by the General Telecommunications Law (Law 9,472, 16 July 1997) and governed by Decree 2,338 of 10 July 1997 [2]. This agency is administratively and financially independent and is not hierarchically subordinate to any government body. Their decisions can only be appealed in court. ANATEL inherited from the Ministry of Communications the powers to grant, regulate and supervise telecommunications in Brazil, as well as a large part of the technical expertise and other material assets.

3.9 DETECTIONS OF ELECTROMAGNETIC EFFECTS WITHIN EVERYONE'S REACH

In HBC, the impact of the antenna on the human body is poorly understood and, consequently, there are uncertainties regarding signal propagation and interference mechanisms due to the effects of the antenna on the human body. Furthermore, some experimental phenomena are not yet well explained in the literature. For example, there is no analysis of why the maximum HBC gain occurs around 50 MHz. However, any citizens can easily observe for themselves the varied effects that can be caused to the environment by electromagnetic waves using objects available at home, such as [18–20]:

- Microwave ovens (2450 MHz, corresponding to the vibration frequency of water molecules);
- Short-circuit incandescent lamps that light up when approached within 3 m of an antenna transmitting any commercial electromagnetic radiation;
- Many LED lamps; most manufacturers operate their products with frequencies above the 100 Hz range, typically above 250 Hz;

- The human body, which is a conductor of electricity (as it is more or less 60% salt and water) and, therefore, is subject to the induction of electrical currents and the effects of electromagnetic waves, as with any other conductor;
- Cell phones and the human ear, which can be heated via calls (>3 minutes).

The idea that cell phones have become consolidated in society as something far beyond a means of communication is widespread, as they are used to carry out the most diverse social, professional, security, educational activities, etc. A cell phone can contain more than 30 functions, including: telephone, calculator, recorder, camcorder, camera, news channels, clock, timer, calendar, diary, internet access, alarm clock, reminder, dictionary, maps and many others. To achieve all this, the electromagnetic signal received and emitted by these devices is made up of similar waves to those of the signal emitted by radio and television stations. This is the least energetic radiation in the entire electromagnetic spectrum, and is also considered non-ionizing radiation; that is, it is not capable of producing appreciable effects on cells.

The main aspects of how the legal control of electromagnetic propagation occurs in the modern world can be observed as follows:

1) Large cities in the world do not have cell phone antennas (survey carried out using Google Earth);
2) Interested in cell phones (points of view):
   - Telecommunications companies (commercial profits);
   - Consumers (safe and healthy environment);
   - Politicians and public bodies (resources for campaigns, jobs, increased revenue and consumer votes);
3) The installation of towers invades the privacy of neighboring properties since nothing can be built within the minimum distance stipulated by municipal laws and decrees;
4) A tower of the size used by base radio antennas will certainly:
   - Change the look of the surroundings;
   - Expose residents to radiation whose effects on humans are uncertain;
   - Ostensibly impose its presence indiscriminately on everyone regardless of consultation, except the owner of the area where it will be installed;
   - Affect ecology when the antenna is installed on the outskirts of cities (tree felling, death of birds by burning, sterilization of people, animals and, possibly, plants);
   - Be imposed indiscriminately on the entire surrounding area; cell phone use is an individual decision and the base radio tower is not.
5) Most publications say that radiofrequency radiation (RFR) is harmless, but they always end up suggesting, without any emphasis, precautions such as:
   - Use the cell phone 2.5 cm from the ear;
   - Global practice of avoiding cell phone use by children;
   - Use the device's speaker or headset;
   - Use the cell phone alternately on the left and right ear;

- If a pacemaker is used, always use the cell phone in the ear opposite to where the pacemaker is installed;
- Avoid installing a radio base station (RBS) close to hospitals, daycare centers, geriatric clinics, public places and schools.

6) Absorbed levels of electromagnetic energy cause temperature increases of 1–2 °C in living beings, and cause a large number of physiological effects characterized in studies with cellular and animal systems. These effects include [21–24]:
   - Changes in neural and neuromuscular functions;
   - Increase in the permeability of the blood–brain barrier;
   - Lack of eye control (lens opacity and corneal abnormalities);
   - Changes in the immune system associated with fatigue and insomnia;
   - Hematological changes;
   - Reproductive changes (e.g. reduced sperm production);
   - Teratogenicity;
   - Changes in cell morphology, water and electrolyte content and membrane functions.

7) The laws and decrees are suggested by the telecommunications companies themselves, interested in the issue and possessing knowledge, measuring equipment, experience in dealing with public bodies and lobbies, which does not happen with the isolated consumer; as far as we know, no researcher or professor or other science and technology (S&T) body was consulted when drafting municipal codes;

8) Disproportionality in dealings between consumers and telecommunications companies that already have globally tested routines in case there are community movements against their products, drastically reducing the chances of success in localized reactions;

9) The sharp real estate devaluation of properties surrounding the installation sites of base radio towers;

10) How can it be explained that some laws include 'sensitive areas', in countries like Brazil (Law 8,896 of 26/04/02 of Porto Alegre and Law 4444/01 of 15/08/01 of Santa Maria), assuming that in its surroundings there are no people who depend on medical equipment to support themselves, there are no children, there are no people with similar characteristics to those in hospitals, etc., who do not benefit from the same precautions as hospitals, daycare centers, schools and geriatric clinics?

11) Why are only the use of medicines and drugs and, more recently, tobacco and pesticides, so well tested before going to the consumer?

An interesting conclusion at this point is the possibility that organs of living beings are affected by intermediate frequency bands since the eyes can usually sense frequencies ranging from the color red at 405 THz (1 THz = 10^{12} Hz) to the color violet (up to 790 THz), while electromagnetic waves from communication systems only range from 900–1800 MHz (1 MHz = 10^6 Hz), see Chapter 2. These aspects of electromagnetic waves are discussed in more detail in Chapter 9 [21].

3.10 ASPECTS OF PUBLIC HEALTH

It is very difficult in the modern world to have reasonable control without excessive limitations for the media or public and private bodies. However, some aspects can be well considered with regard to the health of living beings [22–25]:

1) Demand and publish technological and scientific studies on the impacts of antennas on the neighborhood and ecology;
2) Monitor the community evolution of diseases and changes such as cancer, vascular diseases, adverse reproductive outcomes (sterility, congenital malformations, mental illnesses and others), eye diseases (cataracts, blindness, retinal detachment and others), heart diseases, leukemia, brain tumors, stress, insomnia and others using existing data from hospitals, clinics and health departments; use the databases of health secretaries, hospitals and other health bodies to prepare studies and temporal evolution graphs of diseases;
3) Require public and well-advanced disclosure, mainly to the neighborhood, of tower and antenna design data;
4) Interact in cooperation and agreements with institutions and public bodies related to various environmental specialties;
5) Manage with public bodies the reduction of maximum radiation levels to 1 W/m²;
6) Require the deactivation of antennas already installed too close to communities, hospitals, schools and daycare centers, mirroring examples from Italy, Switzerland, Porto Alegre-RS-Br and Campinas-SP-Br;
7) Require that the information provided to the owner of the area leased to the ERB be from entities exempt from interests other than those of a healthy environment (MO'Ã, CONDEMA, IBAMA, University Advisory Groups, others);
8) Oblige the granting public authority or telecommunications companies to publish regular measurements of electromagnetic power density in all locations where there may be access by the general population;
9) Telecommunications companies must always be given a deadline to clarify doubts regarding controversies, the methodology and techniques used in publications on electromagnetic radiation regarding harmful effects on health or, at least, which companies' projects are intended to contest or address these doubts for the population.

REFERENCES

[1] H.H. Skilling, Fundamentals of Electric Waves, 2nd ed. John Wiley and Sons and Toppan Company, 1948.
[2] A. Zangwill, Modern Electrodynamics. Cambridge University Press, ISBN-10: 0521896975, ISBN-13: 978-0521896979, 31 Dec 2012.
[3] Regulamento sobre Limitação da Exposição a Campos Elétricos, Magnéticos e Eletromagnéticos na Faixa de Radiofrequências entre 9 kHz e 300 GHz (Regulation on Limiting Exposure to Electric, Magnetic and Electromagnetic Fields in the Radio Frequency Band between 9 kHz and 300 GHz), Resolution ANATEL (Agência Nacional de Telecomunicações), 303, 02 Jul 2002.

[4] W.M. Leitão Tavares, Radiação de antenas do serviço móvel celular e seu tratamento na legislação brasileira e de outros países. Digital Library of the Chamber of Deputies, Documentation and Information Center, Library Coordination, http://bd.camara.gov.br, 2004.

[5] P. Knipe and P. Jennings, Electromagnetic radiation emissions from RAPS equipment. In 42nd Annual Conference of the Australian and New Zealand Solar Energy Society (ANZSES 2004), Perth, Western Australia, 30 Nov 2004.

[6] Anne Brice interview with Joel Moskowitz, researcher in the School of Public Health and director of the Center for Family and Community Health at UC Berkeley, Berkeley News, https://news.berkeley.edu/2021/07/01/health-risks-of-cell-phone-radiation, 01 Jul 2021.

[7] A.C. Marconi Stipp, S.A. Tovo Abud, and J.E.R. Duran, Effects of the radiofrequency radiation on submandibular gland of rats foetus, Magazine FOB, 4(314), 27–31, Jul–Dec 1996.

[8] A. Molteni, J.E. Moulder, E.P. Cohen, B.L. Fish, J.M. Taylor, P.A. Veno, L.F. Wolfe, and W.F. Ward, Prevention of radiation-induced nephropathy and fibrosis in a model of bone marrow transplant by an angiotensin II receptor blocker, Experimental Biology and Medicine (National Library of Medicine, National Center for Biotechnology Information), 226(11), 1016–1023. https://doi.org/10.1177/153537020122601108, PMID: 11743137, Dec 2001.

[9] ICNIRP, RF EMF GUIDELINES 2020, Guidelines for limiting exposure to electromagnetic fields, 100 kHz to 300 GHz, Health Physics, 118(5), 483–524. https://doi.org/10.1097/HP.0000000000001210, 2020.

[10] Resolution of notice of inquiry, second report and order, notice of proposed rulemaking, and memorandum opinion and order, Commissioner Rosenworcel concurring, Federal Communications Commission FCC 19-126, Before the Federal Communications Commission Washington, D.C. 20554, 04 Dec 2019.

[11] B.M. Kibret, The Human Body Antenna: Characteristics and Its Application, PhD thesis, College of Engineering and Science, Victoria University, Melbourne, Australia, Jan 2016.

[12] S.R. Chowdhury and K. Ali, Effects of human body on antenna performance: A quantitative study. In Proceedings of the International Conference on Computer and Information Technology (ICCIT), Florence, Italy, 11–12. IEEE, pp. 108–112, Apr 2016.

[13] L. Vijayalakshmi and P. Nirmala Devi, Impacts of RF radiation from mobile phones on human health and its remedies, Journal of Applied Research and Technology, 18(5), ISSN: 2448-6736 (on-line), ISSN: 1665-6423 (Print), Department of Electronics and Communication Engineering, Kongu Engineering College, Perundurai, Erode – 638060, Ciudad de Mexico, https://doi.org/10.22201/icat.24486736e.2020.18.5.1282, Oct 2020 Epub, 30 Jul 2021.

[14] N. Parhizgar and M. Lak, Effect of presence of human body on antenna gain Asma Lak1, Indian Journal of Science and Technology, 8(30). https://doi.org/10.17485/ijst/2015/v8i30/74568, ISSN: 0974-6846 (Print), ISSN: 0974-5645 (Online), Nov 2015.

[15] D.I. Mcree, Soviet and Eastern European research on biological effects of microwave radiation, Proceedings IEEE, 68(1), 84–91, Jan 1980.

[16] V. Popova and A. Shevchenkoa, Analysis of standards and norms of electromagnetic irradiation levels in wireless communication systems on railway transport, Elsevier, ScienceDirect, ICTE in Transportation and Logistics 2018 (ICTE 2018), Procedia Computer Science, 149, 239–245, 2019.

[17] H. Eger, K.U. Hagen, B. Lucas, and H. Voit, The influence of being physically near to a cell phone transmission mast on the incidence of cancer, Journal of the National Cancer Institute (JNCI), 17, 1–7, Jan 2004.

[18] A. Balmori, Evidence for a health risk by RF on humans living around mobile phone base stations: From radiofrequency sickness to cancer, Environmental Research, Science Direct (Elsevier), 214(part 2), 113851. https://doi.org/10.1016/j.env res.2022.113851, Nov 2022.

[19] R. Keller, Electromagnetic Radiation and Adverse Health Effects. Academy of EMC, Education for Professional Engineers, www.academyofemc.com/post/electromagne tic-radiation-and-adverse-health-effects, 23 Dec 2022.

[20] A. Balmori, Evidence for a health risk by RF on humans living around mobile phone base stations: From radiofrequency sickness to cancer, Environmental Research, Elsevier, PMID: 35843283, DOI: 10.1016/j.envres.2022.113851, Vol 214, Part 2, Nov 2022.

[21] F. Soler, Radiation Effects of Wearable Antenna in Human Body Tissues, Dr. Heather Song, Director, University of Colorado Springs, 2014, pp 1–62.

[22] O.P. Gandi, Gianluca L., and C.M. Furse, Electromagnetic absorption in the human head and neck for mobile telephones at 835 and 1900 MHz, IEEE Transactions on Microwave Theory and Techniques, 44, pp 1884–1897, retrieved Oct 1996.

[23] N.H. Abd Rahman, Y. Yamada, and M.S.A. Nordin, Analysis on the effects of the human body on the performance of electro-textile antennas for wearable monitoring and tracking application, Materials (Basel), 12(10), 1636. Published online 19 May 2019. https://doi.org/10.3390/ma12101636, PMCID: PMC6567044, PMID: 31109128, May 2019.

[24] P.A. Hasgall, F.D. Gennaro, C. Baumgartner, E. Neufeld, B. Lloyd, M.C. Gosselin, D. Payne, A. Klingenböck, and N. Kuster, Data from: IT'IS Database for Thermal and Electromagnetic Parameters of Biological Tissues V4 [Dataset]. IT'IS Foundation, 2018. Accessed on 28 Dec 2018. Available online.

[25] S.J. Boyes, P.J. Soh, Y. Huang, G.A.E. Vandenbosch, and N. Khiabani, Measurement and performance of textile antenna efficiency on a human body in a reverberation chamber, IEEE Transactions on Antennas and Propagation, 61(2), 871–881. https://doi.org/10.1109/TAP.2012.2225817, INSPEC Accession n. 13287835, 19 Oct 2012.

4 Information Storage in the Brain

4.1 INTRODUCTION

There is a close connection between the brain and the vibrations perceived by the various organs of living beings, such as those arising from sounds, colors and touches on the skin, which are made through neurons or nerve cells. Neurons emit chemical and electrical signals involved in this short-distance communication with nearby cells. An electrical current is thus formed that can carry out long-distance body–brain communication. This long-distance path refers to the signal going from the muscles and other parts of the body to the brain and vice versa. As an electric current follows such a winding path, there is the expectation of possible external electromagnetic interference caused by the electromagnetic communications means, as discussed in Chapter 3. This indicates that the body must react to the received signal in some way, for example to blink an eye, take a step, move the arms, use the hands for some action, run away from some danger, and other similar functions [1]. In this chapter, it is shown how all this happens.

4.2 PERCEPTION OF VIBRATIONS BY THE BODIES OF LIVING BEINGS

It is known that electrochemical brain waves, like any others, have well-defined characteristics that are dependent on the length of the vibrational wave, λ, which has an inverse relationship with its frequency, f, and this, in turn, defines the number of repetitions of a certain periodic phenomenon. One can thus obtain the wavelength using its propagation speed, v, times its repetition period, T, or its speed divided by the signal frequency, equation 4.1.

$$\lambda = vT = \frac{V}{f} \tag{4.1}$$

After the signals received by the sensory organs have reached the brain, they are stored or identified in some way with what is already recorded there throughout a person's life experience and thus generate the appropriate bodily reactions.

DOI: 10.1201/9781003604037-4

4.3   IN-MEMORY DATA STORAGE

Data storage in human memory occurs through neurons, which are the cells of the nervous system that transmit nerve impulses. Although they are not the only cells in nervous tissue to perform this function, neurons stand out as being the best known cells for this.

Neurons have the following basic parts: cell body, dendrites, axons and synapses. The cell body is the region where most of the neuron organelles are located and, therefore, a cell can receive nerve impulses from other nerve cells there through the dendrites that are extensions of this cell [2].

It is known that neurons communicate with each other at points of contact called synapses, sending a message to a target neuron that is nothing more than another cell. Most synapses are chemical and communication with them is done using chemical messengers. Other synapses are electrical, which is the path through which a direct flow of ions occurs between cells. At a chemical synapse, an electrical action potential causes the presynaptic neuron to release information neurotransmitters. These molecules bind to receptors on the postsynaptic cell and make it more or less likely to trigger an action potential.

Axons are extensions that are much longer than the cell's dendrites and are generally solely responsible for transmitting nerve impulses to other cells in the body. The axon's signals are generated at a cone-shaped end close to the cell body, called the implementing cone. At its other end, several branches transmit signals to other cells. This junction region is called the synapse and the branch of this junction is called the synaptic terminal.

The Schwann cell, which is responsible for the synapse connections, is the unit of structure, physiology and organization of living beings. Theodor Schwann (1810–1882), a German physician and physiologist who is considered to be the founder of modern histology, developed the basis of cellular theory, which is used to describe the elementary anatomical composition of plants and animals [3].

A synapse is a specialized junction between a neuron and a target cell, the region where nerve impulses are transmitted from one cell to another. The branch of this junction is called the synaptic terminal. The synapse is the region of the neuron that carries out communication between two or more neurons, or from a neuron to an effector organ, that is, to a muscle or a gland. These synapses send signals through synaptic transmission to check some specific action that has been carried out by the body. During this synapse action, the neuron that transmits the impulse is called a presynaptic cell and the cell that receives the impulse is called a postsynaptic cell, which can be another neuron, or a gland, or even a muscle cell.

Myelin is a plasma membrane characteristic of special cells that surround the axon of certain neurons. The myelin sheath is a structure formed by a lipid membrane rich in glycophospholipids and cholesterol, which covers the axons, as shown in Figure 4.1. These sheaths act mainly as electrical insulators, facilitating rapid communication between neurons. The myelin sheath is formed by glial cells that are specialized in the most diverse functions, including protecting and nourishing neurons. Glial cells are non-neuronal cells of the central nervous system interacting with neurons and are responsible for supporting and nourishing neurons, keeping them together in addition

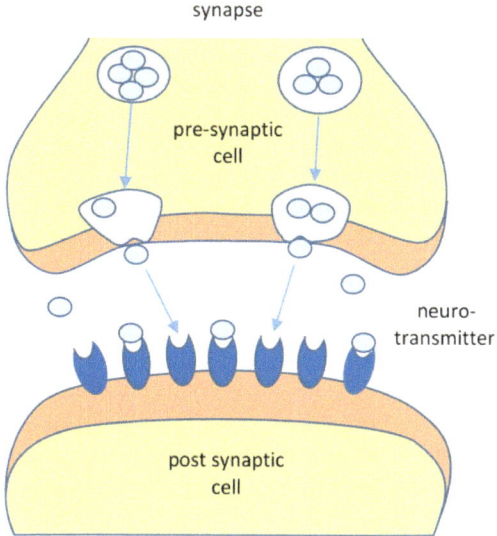

FIGURE 4.1   Neuro-data transmission.

to other important activities. Glia with an estimated ten cells for each neuron are much more numerous than neurons. When a person is asleep and has more vivid dreams, brain activity works in the same way as when this human being is awake.

4.3.1   EXPLICIT MEMORY

Explicit memory is made up of the basal ganglia and cerebellum, which are the basal ganglia. These structures are situated deep within the brain and are involved in a wide range of processes, such as emotions, reward processing, habit formation, movement and learning. They are particularly involved in coordinating sequences of motor activity, as would be necessary when acting out in activities such as a theatrical scene, dancing, playing a musical instrument, playing football or swimming. The basal ganglia are the regions most affected by Parkinson's disease, with patients having impaired movement of their external organs. This explicit memory is composed of the hippocampus, neocortex and amygdala, each with a very explicit function, as follows [4–6]:

- **Hippocampus**
  The hippocampus is the part of the brain where episodic memories are formed and indexed for later access. Such episodic memories are autobiographical for specific events in human lives, such as drinking in a bar with a friend last month, parking a car in a busy road, fighting with a spouse or partner, or a setback of some kind.
- **Neocortex**
  The neocortex is a tissue that forms the outer surface of the brain. Over time, information from certain memories that has been stored temporarily in the hippocampus – for example, knowing to prepare for a professional test – can be

transferred to the neocortex as general knowledge. A large number of researchers think that this transfer from the hippocampus to the neocortex happens during sleep.

- **Amygdala**

  The amygdala assigns emotional meaning to memories. This is particularly important because strong emotional memories are difficult to forget. For example, those memories associated with accidents, shame, pain, joy, fear, love, nice times or sadness. The permanence of these memories suggests that interactions between the amygdala, hippocampus and neocortex are crucial in determining the 'stability' or effectiveness of how a memory is retained over time.

4.3.2   IMPLICIT MEMORY

Information that people do not remember on purpose is stored in implicit memory or in the unconscious or automatic memory. This type of memory is unconscious and unintentional and is therefore sometimes called non-declarative memory, since the person is not able to consciously bring it to consciousness. Unlike explicit memories that are conscious and can be explained verbally, implicit memories are generally not conscious and cannot be articulated verbally. These memories are much more focused on what is going to be done and the step-by-step processes that must be carried out to complete an already well-known task.

Procedural memories refer to specific tasks, such as calling a lift, keying a door, turning on a lamp, driving a car or going for a walk. These are implicit memories, as the author does not need to be consciously thinking about how he/she is going to perform these tasks. Even though implicit memories are not consciously remembered, they are the basis of human behavior and related tasks [7–9].

4.3.3   MEMORY IN ACTIVITY

The active memory is the prefrontal cortex (PFC), which is the part of the neocortex that is right at the front of the brain. This memory is the most recent addition to the mammalian brain and is involved in many complex cognitive functions. Human neuroimaging studies obtained with magnetic resonance imaging (MRI) machines show that when people perform tasks that require storing information in their short-term memory, the PFC becomes active. This is the case for the location of a light flash or the fast approach of a car when moving from one sidewalk to another. There also appears to be a functional separation between the left and the right sides of the PFC. The left side is more involved in verbal working memory, while the right side is more active in spatial working memory, such as remembering where the light flash occurred or whether there will be time to cross the street. All of this is trained and accumulated throughout the experiences of each individual's life.

4.3.4   MEMORY FORMATION

Memories are constantly formed in the brain by different groups of neurons, conducting different thoughts or perceptions, coming and going in action according to the need already perceived in other experiences. In other words, they are a reactivation of a specific group of neurons. However, what allows a particular combination of neurons to be reactivated relative to any other combination of neurons is synaptic plasticity. This term describes persistent changes in the ability of synapses to connect between brain cells. These connections can be more or less deep, depending on when and how often they were activated in the past. Active connections tend to become stronger, while those that are not used become weaker and may eventually disappear completely over time [10,11].

Few features of the brain are more important than synaptic plasticity. Changing the strength of existing synapses, or even adding new ones or removing old ones, is fundamental to memory formation. However, there is also evidence that there is another type of plasticity, which does not directly involve synapses, and which could be important for memory formation. New neurons can be created in some parts of the adult brain in a process called neurogenesis, such as an important memory structure known as the hippocampus. Studies with older rats have shown that by increasing neurogenesis in the hippocampus, memory can be improved. In humans, memory use has been shown to increase the volume of the hippocampus, suggesting that new neurons are being created. This more intense use of memory at the same time improves performance in tasks that rely more heavily on memory.

4.3.5 PATHS OF MEMORY

An acceptable analogy for memory formation is the way foot traffic creates a path on a lawn. The more a piece of grass is trampled, the more defined the path becomes and therefore, the easier it is to follow it as if the 'memory' of the walk had been established. The same thing happens in the brain. The more a neural pathway is activated, the stronger the synaptic connections along the path become. So, when a thought enters the brain, for example, a very pleasant event, the related experiences or knowledge are easily recalled. This happens with nice relationships, parties, smiles, dangers, pleasant or unpleasant scenes, or as the mind channels these thoughts along previously well-established neural pathways [12,13].

4.4 COMMUNICATION BETWEEN NEURONS

Modulation of synaptic connectivity is a critical learning mechanism. This explanation was incorporated by Hebb in the 1940s and 1950s to be used in more refined models [14] to explain mechanisms involving nerve impulses, which are connections established between one neuron and others. The nerve impulse traveling through the neuron is electrochemical in nature and results from internal and external modifications of the neural membrane. Internally, the neuron membrane has a negative electrical charge, while externally it is positive. Therefore, the nerve impulse corresponds to an electrical current that propagates through neurons thanks to a process of membrane depolarization that occurs upon encountering stimuli. This connection has the function of transmitting signals to other nerve cells, whether muscular or glandular.

According to Figure 4.1, it can be observed that synapses or connections between neurons can occur in two ways. One is through electrical synapses: when electrical current flows from one neuron to another through gap junctions that form cytoplasmic channels to allow cell-to-cell communication, common in invertebrates; and the other is through chemical synapses: information is transmitted from one neuron to another through chemical interrelationships called neurotransmitters. The presynaptic neuron synthesizes and packages neurotransmitters into so-called synaptic vesicles. With the arrival of the electrical impulse and the depolarization of the presynaptic membrane, these neurotransmitters are released to stimulate a nerve.

Certain synapses on recruited neurons are more prone to increased synaptic strength, a phenomenon known as long-term potentiation (LTP). Mechanisms that may contribute to allocation at the synaptic level include synaptic tagging, capture and clustering.

The synapse is an important functional part of the brain. Over the past few decades, scientists have gained a wealth of information about its structure, molecular components and physiological function. In the 1940s, after World War II, Norbert Wiener helped the Massachusetts Institute of Technology (MIT) recruit a group of cognitive science and researchers in neuropsychology, mathematics and the biophysics of the nervous system. Warren Sturgis McCulloch and Walter Pitts joined MIT and became the fathers of artificial neural networks. They made pioneering contributions to computer science and artificial intelligence. McCulloch and Pitts' contributions are the beginning of two fields: the theory of finite state machines as a model of computation, as well as the field of artificial neural networks, which has been renamed in recent years as deep learning.

It is clear that synapses are different from each other morphologically and molecularly and are recruited for different functions. One of the most intriguing findings is that the size of the synaptic response is variable and can be altered by a variety of factors, such as the homo- and heterosynaptic, which are prior patterns of use or modulatory neurotransmitters. Perhaps the most difficult challenge for neuroscience is to perform experiments that can better clarify these basic building blocks of the brain structure. These blocks are assembled and regulated by the flow of information through the neural circuits. From that, it produces complex behaviors in such a way as to be able to store them in an organized way in memories.

4.5　HOW DATA APPEAR IN MEMORY

Dreams can manifest impulses from the body core, remote memories and associations of ideas or simple random communication between neurons. The impulses of thoughts or the human core cause random actions in some part of the body resulting from the accumulation of each person's daily experiences and the excess force or electricity necessary for their activities. They can also signal that something is wrong with the living person, but they can also just be involuntary impulses from the brain to release electrochemical excesses arising from information received throughout life and an exaggerated accumulation of it.

The manifestation of remote memories about the pleasurable effects of premarital sex, drugs, games, victories or alcohol play an important role in the relapse of potentially addictive behaviors. Certain triggers, such as alcohol advertisements, naked bodies or walking past a favorite bar, can evoke memories that make it very difficult to avoid temptation and not fall back into addiction, even if these people have long ago stopped drinking, games, drugs or getting involved in sexual adventures [15].

The association of ideas may be present when two phenomena have often been experienced in conjunction, but have not occurred separately in any isolated instance. In other words, in experience or thought what has been called an inseparable association occurs between them. Less correctly, it is known as an indissoluble association, as this association cannot be said to last inevitably until the end of life. With this, no subsequent experience or thought process could be helpful in dissolving it, and just as long as such an experience or thought process has not occurred, the association is irresistible; it is practically impossible to think of one thing apart from another [16,17].

4.6 TRANSMISSION OF A NERVE IMPULSE

Another important aspect to understand about dreams is what in neurophysiology, or more specifically in neuroscience, is metaphorically called closets, deposits or just a simple random neuronal communication.

4.6.1 OPTICAL COMMUNICATION MECHANISMS

The retina, vitreous and optic nerve are the main mechanisms of visual communication between the environment and the brain of animals, including humans. The retina is the membrane of the posterior segment of the eye, highly vascularized, composed of ten layers of nervous tissues. The central region of the retina is called the macula and is responsible for visual details. The retina is one of the membranes in the posterior segment of the eye with the function of transforming the light stimulus into a nervous stimulus and, from there, sending it to the brain through the optic nerve so that the images are processed, compared and recognized as memories, previously perceived or not.

The vitreous is a gelatinous substance that fills the inner and posterior part of the eye cavity that is in contact with the retina surface. As a person ages, the vitreous undergoes a process of liquefaction and condensation of its fibers, resulting in separation from the retina. When this occurs, the symptoms are light flashes and the sudden onset of floating spots that look like tiny insects flying around. To do this, the patient must be examined by an ophthalmologist specializing in the retina and vitreous and define the problem exactly.

The vitreous is a transparent tissue that fills the ocular cavity between the lens and the retina at the back of the eye, consisting mainly of water and a substance called hyaluronate, as well as collagen and other proteins [18]. The function of the vitreous is to help maintain the shape of the eye and support the retina.

4.6.2 ITEMS STORED IN BRAIN CABINETS

Imagine that a person is in a hurry to go out, starts looking for a certain pair of gloves in a closet, and only finds one of the gloves. Then, in a hurry, this person opens several other places where the missing glove could be and, not finding it, this person takes another one, of approximately the same color, sometimes realizing this only after some time. As the colors are approximate and the coat will almost completely cover the gloves, people in a hurry use different pairs of gloves, and this decision may be conscious or not. Another, somewhat dramatic, example could be the case of a person going to an emotional event, such as to visit someone in a bad state, a burial or the trial of a crime that affects them. It may happen that in a hurry this person wears a shirt inside out, only realizing this when other people at the same ceremony look at this person in surprise.

In a state of alertness, the brain can be distracted, especially when a person is sleeping. Suppose a mosquito has bitten this person's face. The epidermal sensory reaction may be similar to that of a bump caused by a pimple. As the brain works to understand everything as if it were stored in a closet, this person may have a dream in which a pimple begins to develop on the face. This bulge, in turn, can lead the brain to something else it considers similar. Therefore, what first seemed like a pimple in the dream can turn into a finger, depending on the person's imagination and from there, start a new brain search for something similar. Then the brain, once again, starts searching in one of the mental closets for what seemed to be missing and then it can find a knife. This may have nothing to do with feelings, repressed sexual desires or repressed desire for power, but rather with a neuronal capacity to try to compose scenarios through approximation or mnemonic association.

Any dream can be the result of what is stored in the brain, the result of information coming from the gathering of images, sounds and associative sensations that accumulate in the brain throughout life, even before birth. In addition, natural neurophysiological transformations occur that are related to internal and external stimuli. There are dreams most of the time with real meanings and some that can even anticipate facts in the person's life through an unconscious association with what has previously been experienced or imagined. With this, the brain accesses a network of information and, through ordinary reasoning, creates new estimates with feasible possibilities that are more or less correct. However, most dreams may just be a composition, a random mnemonic approximation, an exaggerated imagination, or a personal desire.

Dream analysis should be performed by professionals with high competence and knowledge of how to analyze their meanings, always taking into account the most important aspects of a brain functioning in relation to the entire physiological context of the human being involved. The only way to achieve this is always the association with the cognition origins in the patient's human evolution without creating expectations that only a fanciful imagination would explain. For this, it is necessary to remember that neurosciences in their development and abstraction have revealed many new aspects of brain functioning, making it possible to review the concepts proposed by Jung, Freud, Karl Abraham and many others [17]. Although Freud and Jung were doctors, what they did was not exactly neuroscience, but rather the beginning of modern psychoanalysis and psychotherapy, where the ancient 'alienist' effectively became a therapist. In Jung's case, he distanced himself even further from Freud by basing his life and career on studies of human personality. Jung, whose studies were continued by many other researchers, proposed 'common' definitions of ego, superego, shadow work and other notions.

4.6.3   FREE ASSOCIATION AND DREAM INTERPRETATION: A CASE STUDY

In the history of psychology, Freud used the method of Antiphon, born around 480 BC, and Artemidorus de Daldis, who lived in the second half of the second century, to delve into the unconsciousness of his patients. It is about 'interpreting dreams', not exactly through the elements presented in them, but through the free association of ideas that the patient makes in relation to these elements. Of course, there is a factual deficiency in this method that must be carefully observed by the analyst since free associations may not, in fact, be free. This means that the patient, without realizing it, tends to 'run away from the truth', which, perhaps, he or she is not ready to face. With this, this patient carries out what is called 'psycho-associative defense'. Even so, in general, this method is quite efficient.

To demonstrate how complex dream interpretation is, imagine that a person has a recurring dream and that, in this dream, this person always kills and dismembers children and tries to hide the crime. In the dream, such attitudes that are repeated with little change in scenery and elements will always be devoid of pleasure, anger or hatred feelings. There is total apathy, an anesthetization broken only by the fear of punishment, the police or someone else being able to find out, all depending on how strong or weak the participation of the person involved in these dreams is.

4.7   DATA ALLOCATION IN MEMORY

Data allocation in memory is the process that determines which specific synapses and neurons will store a given memory in a neural network. Although multiple neurons can receive a stimulus, only a certain subset of neurons will induce the plasticity necessary for memory encoding. The selection of this subset is called neuronal allocation. Likewise, multiple synapses can be activated by a certain set of inputs, but only specific mechanisms determine which synapses will actually encode the memory, a process known as synaptic allocation.

Sheena Josselyn and colleagues in the laboratory of Alcino J. Silva [19] first visualized data allocation in memory in the lateral amygdala. At the neuronal level, cells with higher levels of excitability, such as those slower after hyperpolarization, become more likely to be recruited by a memory trace. There is substantial evidence that involves the cellular transcription factor in this process known as responsive element binding protein, cyclic adenosine monophosphate (cyclic AMP) or cAMP-response element binding protein (CREB).

There are many speculations about how a human being can completely forget specific information or dreams. A new study carried out in Japan by researchers from several medical entities may have achieved an acceptable answer for this. A publication from this new study indicates that the human brain can force the loss of memories and that a set of neurons exists exclusively to eliminate memories considered irrelevant, causing the individual to forget small details of everyday life. The brain selects the information that will be stored and, during the day, it is common for people's minds to become too full of generalized information that, at the end of the day, is useless. Therefore, brain cells, known as neurons, that house melanin-concentrating hormone (MCH) (protein), actively destroy these irrelevant memories during rapid eye movement (REM) sleep.

It is estimated that the brain has a capacity of 100 terabytes (TB) of memory and that human consciousness can only access more or less 10% of it (1 TB = 1 tera-byte = 10^{12} bytes, 1 byte = 8 bits and a bit can be 1 or 0, that is, on or off, respect-ively). This number gives the impression that every human being's brain has infinite capacity. Even with all this, many have difficulty memorizing a simple situation, text, name, birthday, telephone number or computer link. Neuroscientists have long been trying to measure how much more exactly can fit into a human memory, but the task becomes almost impossible when countless cases of people with very few skills are known of, alongside others who are extremely dedicated and perform incredible feats with their brains. One of them is Chao Lu, who in 2005, when he was still a 24-year-old university student, managed to correctly recite something around the first 67,890 digits of the number Pi (π), during a period of 24 hours, without breaks [20]. Other people, sometimes considered geniuses, recite numbers in sequence, resorting to another common tactic: converting short series of numbers into words that are joined together by a story. These people perform incredible feats, such as those shown in more details in Chapter 10 of this book, which resemble even the smallest details of an image, like the example given in [21]. This probably has much to do with elec-tromagnetic ambient communications.

In very rare cases, a physical or mental injury can also cause so-called acquired wisdom syndrome. This is what happened to Orlando Serrell, who at the age of ten was hit by a baseball on the left side of the head. Suddenly, he began to show the ability to remember countless vehicle registration plates or even make calculations on many dates from previous decades [22].

The physical volume of adult brains does not differ as much as the impression of their mnemotechnical capacity or density. An example of this is that of Nelson Dellis, champion of the United States Memory Tournament, whose results were from a lot of practice [23]. He said, "I was pretty forgetful, but after a few weeks of training, I found myself doing something that seemed almost impossible. We all have this ability", says Dellis. Like other champions, he uses tried and tested strategies and techniques to memorize items quickly. One of the most used tricks for this is the con-struction of a 'memory palace'. The technique consists of visualizing a place that the individual knows quite well, such as, for example, the house where this individual lived as a child or a public playground. Dellis, for example, then 'translates' the items a person needs to memorize into images by placing them in furniture and corners of the house or on toys in the park. "You mentally navigate that space and select those images that you left there, 'translating' them back into what you memorized", he explains.

The success of the strategies described above indicates that practically anyone can become a memory ace simply by discovering the best own way to memorize, having a lot of dedication or even integrating a mnemonic technique. According to Allen Snyder, director of the Center for the Mind at the University of Sydney, Australia, everyone has an 'inner genius' that can be revealed if the correct technique is applied. Snyder asserts that the human mind primarily operates at a high level of conceptual thinking, rather than focusing on small details. "We are aware of the whole, but not the parts", he explains [24].

It is possible that human evolution helped the brain function in an automatically defensive way. For example, instead of paying attention to every detail when someone walks when crossing a street, the mind warns this person of the danger that could be an approaching car and, in a matter of milliseconds, warns of the type of danger that is faced and must be reacted to immediately. In other words, most of the data that the senses transmit to the brain are not taken to the conscious level, going directly to the subconscious, depending on the case. However, for people considered 'geniuses', conceptual thinking is not as present, giving them time and space to detail every second they experience.

Cases of acquired wisdom syndromes, such as that of the boy Serrell and other cases, and those cited in Chapter 10 of this book, led Snyder to seek a physiological basis for the phenomenon. Snyder believes that the left anterior temporal lobe, just above the left ear, plays the most important part in it. Other researchers have also noted a certain dysfunction in cases of autism and dementia praecox, which can be accompanied by exceptional mathematical, artistic or musical talent (see Chapter10). However, despite the ambitions of scientists like Snyder, it will still take some time to discover a shortcut to becoming a genius. Other factors, such as self-confidence, prejudices, traumas and the level of mind alertness, can influence learning [24,25].

4.8   VARIABILITY OF NEURON DISTRIBUTION

A key problem when trying to count the brain's neurons is the variability of their distribution throughout the brain. Neuron density can vary by factors of up to 1,000 between structures. Even within a single structure, different layers can consist of different numbers of neurons. Therefore, after numerous attempts by scientists to determine this, a method to develop a reliable and valid brain cell count including the count of neurons, non-neuronal brain cells and endothelial cells, known as an 'isotropic fractionator', was developed. The method involves dissolving cell membranes while preserving nuclear membranes, producing what Herculano-Houzel called brain 'soup'. Each neuron consists of a nuclear membrane and in this brain soup there are floating nuclei that are relatively easy to count by sampling small quantities. To do this, all cell nuclei are stained blue, collected and counted. Counting nuclei is simple and does not require special training. In the book, *The Human Advantage*, in 2016 Herculano-Houzel provided a description of what went into developing the technique, including details about the first failed attempts to create a brain soup [26–31]. Other researchers, including Christopher von Bartheld at the University of Reno and Jon Kaas at Vanderbilt University, have shown that this method is fast, reliable and relatively easy to apply [30,32].

For half a century, neuroscientists thought the human brain contained 100 billion nerve cells. The results of the isotropic fractionator indicate that the human brain has an average of 86 billion neurons and 85 billion non-neuronal cells, which are glial cells and endothelial cells. This figure is countered by Herculano-Houzel, who says that the entire brain of a baboon contains 11 billion neurons and that the difference of 14 billion is a significant number of neurons. It must be taken into account, however, that each neuron can connect to thousands of other neurons. Herculano-Houzel's original research and the research influenced by his discoveries has led to changes in our understanding of the brain and new research by counting brain cells in a variety of different species. Therefore, this work is also important in relation to non-human animals. High-ranking scientific journals have resisted these numbers and it is very difficult to make a definitive statement [32,33].

Regarding the size of the human brain, it can be said that it varies within a certain percentage. For example, renowned physicist Albert Einstein's brain weight was 1,230 g. This weight is very similar to the average brain weight of an elderly control group, which is typically 1,219 ± 102.93 g. However, this weight is lower than that of the young control group, which is 1,374.13 ± 111.56 g. This fact suggests that Einstein's brain weight is just consistent with his age [34]. In addition, scientists also found three interesting things in Einstein's brain:

- Abnormal presence of glial or neuroglial cells that are responsible for helping neurons in the transmission of nerve signals;
- High density of neurons in certain parts of the brain;
- A unique structure of convolutions and fissures.

Despite all the divergences described above, the synapse remains a functional unit of the brain. In recent decades, much new information has been obtained about its structure, molecular components and physiological function. Evidently, synapses continue to be morphologically and molecularly diverse and this diversity is recruited for different functions. One of the most intriguing findings is that the size of the synaptic response can be variable, but can be altered by a variety of homo- and heterosynaptic factors, such as previous usage patterns or modulatory neurotransmitters. Perhaps the most difficult challenge in neuroscience is designing experiments that reveal how these basic building blocks of the brain are assembled and how they are regulated to mediate the flow of information through the neural circuits needed to produce complex behaviors and store memories.

Santiago Ramon y Cajal (1894) first suggested that learning results from changes in synapse strength based on insights from his anatomical studies [35]. This modulation of synaptic connectivity is a critical learning mechanism and was incorporated into models refined by Hebb in the 1940s and 1950s. In 1949, Donald Hebb wrote *The Organization of Behavior*, a work that pointed out that neural pathways are strengthened each time they are used, a concept fundamental to the way humans learn [36,37]. If two nerves fire at the same time, he argued, the connection between them is improved. Hebb's law speaks of a kind of 'synaptic muscle building' involving a mechanism for detecting temporal coincidences in neuronal discharges. Experimental investigation of these intriguing conjectures required the development of behavioral systems in which one could examine changes in the neuronal components of a specific behavior during or after modification of that behavior with learning [36,37].

4.9   CAUSES OF SYNAPTIC CHANGES

Synaptic change is caused by different patterns of neuronal activity at critical points within a behavioral circuit. Both increases and decreases in synaptic strength can contribute to behavioral plasticity [36]. Synaptic plasticity has similar temporal and molecular properties to behavioral learning. One can cite as an example the short- and long-term phases, dependent on discrete signaling pathways. Seemingly different forms of learning use similar underlying cellular and molecular mechanisms.

Long-term memory requires transcriptional activity and genes such as cfos, zif268 and arc, which are rapidly and transiently induced by high-frequency neural activity and have been used for many years to map patterns of brain activity in rodents. By providing a genetic readout of neural activity patterns, these genes have basic potential for gaining direct molecular control over ensembles of neurons based on their response to a given experience.

In one study, the cfos promoter was combined with elements of the Training the Emotional Trading (TET) regulatory system in transgenic mice to allow the introduction of a lacZ marker into neurons activated with fear conditioning [35]. The marker provided a lasting record of brain activity during learning that could be compared to activity during recall. There was a partial reactivation of neurons active during learning, and the strength of the recovered memory was correlated with the degree of circuit reactivation. Most importantly, this approach offers an opportunity to introduce any genetically encoded effector molecule into neurons based on their recent activity, providing the potential to study circuits based on the specific memory they encode.

Many forms of synaptic plasticity differ in their induction mechanism, persistence time and synaptic locus. An interesting insight into this question was provided by a theoretical study that addressed the question of how a memory can persist in the face of the flood of synaptic inputs and synaptic plasticity that a neuron experiences throughout an individual's lifetime. Although it was impossible to encode robust memories with a single form of plasticity, multiple forms of plasticity with distinct time scales of induction and persistence were able to produce persistent memory storage [38,39]. A challenge in the future will be to examine how the diversity of plasticity mechanisms can actually cooperate and interact to produce a unified long-lasting and continuous memory storage mechanism. A surprising discovery in these studies was the considerable molecular detail of the mechanisms underlying long-term synaptic plasticity and the importance that such plastic changes play in memory storage across a wide range of species and memory forms. Allied to this is the remarkable degree of memory conservation mechanisms in different regions of the brain within a species and between species widely separated by evolution. However, although it is clear that long-term synaptic plasticity is a fundamental step in memory storage, it is important to note that a simple improvement in the effectiveness of a synapse is not sufficient to store a complex memory. Instead, changes in synaptic function must occur within the context of a set of neurons to produce a specific change in the information flow through a neural circuit. With the recent development of powerful genetic tools, it is anticipated that it will soon be possible to tackle the enormous challenge of visualizing and manipulating such changes in neural circuits.

A second important challenge in the question of synaptic plasticity is understanding how basic memory storage processes are altered with age or disease, including Alzheimer's disease. There is now convincing evidence that defects in memory storage result from pathological alterations in the fundamental mechanisms underlying the long-term synaptic plasticity. Thus, it is critically important to understand in sufficient detail the basic mechanisms of memory storage and the changes that occur in disease to design specific compounds that can be used to restore cognitive function [38–40].

Memory allocation is a process that determines which specific synapses and neurons are part of a neural network storing a given memory [1–3]. Although multiple neurons can receive a stimulus, only a subset of neurons will induce the plasticity necessary for memory encoding. The selection of this subset of neurons is called neuronal allocation. Likewise, multiple synapses can be activated by a given set of inputs, but specific mechanisms determine which synapses actually encode memory. This process is known as synaptic allocation. Sheena Josselyn and colleagues in the laboratory of Alcino J. Silva [19] first discovered memory allocation in the lateral amygdala.

At the neuronal level, cells with higher levels of excitability (e.g., lower slow after hyperpolarization [32]) are more likely to be recruited to a memory trace. For this, there is substantial evidence that involves the cellular transcription factor CREB (cyclic AMP responsive element binding protein) in this process [41–45]. Certain synapses on recruited neurons are more likely to undergo an increase in synaptic strength, known as long-term potentiation (LTP). Proposed mechanisms that may contribute to allocation at the synaptic level include synaptic tagging, capture and clustering [45].

Synaptic activity can generate a synaptic tag, which is a marker that allows the stimulated spine to capture subsequently newly transcribed plasticity molecules such as Arc. Synaptic activity can also involve the translation and transcription machinery. Weak stimulation can create synaptic tags but will not engage the translation and transcription machinery, whereas strong stimulation will create synaptic tags and will also engage the translation and transcription machinery. Newly generated plasticity-related proteins (PRPs) can be captured by any labeled synapse, while unlabeled synapses are not eligible to receive new PRPs. After a certain time, the synapses will lose their imprint and return to their initial state. Furthermore, the supply of new PRPs will run out. Tags and new PRPs must overlap in time to capture such PRPs [46,47].

The synaptic tag is related inversely to the time between inducing stimuli and is considered temporally asymmetric. Furthermore, tagging is also related inversely to the distance between spines, an important spatial property of tagging. On the other hand, confirming the temporal and spatial properties of synaptic tagging, subsequent imaging studies revealed that there are not only temporal constraints but also structural constraints that limit synaptic tagging and capture mechanisms. Overall, these studies demonstrate the complexity of synaptic tagging and capture and provide further insight into exactly how this mechanism occurs [42].

4.10 MECHANISMS THAT LINK MEMORIES OVER TIME

Denise Cai, in the laboratory of Alcino J. Silva, discovered that memory allocation mechanisms are used to connect or link memories over time [48]. In her studies, she demonstrated that a contextual memory triggers CREB activation and subsequent increases in excitability in a subset of hippocampal CA1 neurons, so that a subsequent contextual memory, occurring within about five hours, can be allocated to some of the same CA1 neurons that stored the first contextual memory. Because of this overlap between the CA1 memory engrams for the two contextual memories, recall of one contextual memory triggers retrieval of the second memory. These studies also showed that contextual memory-binding mechanisms are disrupted in the aging brain and that increased excitability in a subset of CA1 neurons reverses these memory-binding deficits. It is very likely that deficiencies in CREB and neuronal excitability in aging brains may explain abnormalities in memory wiring and they could possibly be the source memory-related problems (source amnesia) associated with aging.

In July 2018, in a special issue on "13 Discoveries That Could Change Everything", *Scientific American* highlighted the Silva lab's discovery on memory allocation and linking [48]. The researchers made the discovery by recording the activity of 2,735 individual neurons in 21 neurosurgical patients implanted with brain electrodes for epilepsy monitoring. They took advantage of a phenomenon known as attentional blink, in which people who pay attention to two familiar images in quick succession often fail to notice the second. The experimental setup allowed the researchers to compare directly the neural response to seen and unseen presentations of the same image. As expected, participants in this study often failed to notice the presence of a second target image, especially when it was presented shortly after the first target image. The researchers found that the corresponding neurons fired anyway. However, there was an observable difference in the strength and timing of this neural response.

"Studying the activity of individual neurons in awake, behaving humans was critical to capturing weak but informative signals from individual neurons during unconscious perception, particularly in regions further down the processing stream that are impossible to measure with conventional tools", says Mormann. "We were quite surprised to see that the timing of neuronal responses indicates whether participants reported seeing the image or not" [49].

The findings weigh on theoretical debates about the nature of human consciousness, researchers say. For example, it is not clear whether consciousness is an all-or-nothing phenomenon or a matter of degrees. The researchers say that the observation of neuronal firing occurs in both cases, but in different ways, arguing for consciousness as a more nuanced and graduated phenomenon. Now they would like to explore how the activity of individual neurons in one part of the brain is related to activity in other areas and how these connections relate to conscious perception. This topic still seems to be extremely broad and without definitive solutions [49–51].

What is known for sure is that human memory has an intrinsic limitation [49]. So the question arises, why can we not remember all the information that reaches the five senses? Reber believes that the brain, when interpreting the world around it, simply cannot keep up with the torrent of external stimuli to which it is exposed. "There is a bottleneck coming from our senses to our memory", he says. Making an analogy with a computer, Reber says that the limit of human memory during a lifetime is not the space on the hard drive, but rather the download speed of this information, since "it happens faster than our memory system is capable of to write them down", he concludes.

REFERENCES

[1] D.R. Hawkins, Letting Go: The Pathway of Surrender, 1st ed. Hay House, ISBN-13: 978-1401945015, 15 Jan 2014.

[2] S.J. Cragg and S.A. Greenfield, Differential autoreceptor control of somatodendritic and axon terminal dopamine release in substantia nigra, ventral tegmental area, and Striatum, Journal of Neuroscience, 17(15), 5738–5746. PMID 9221772, 1997.

[3] T.C. Südhof, The presynaptic active zone, Neuron, 75(1), 11–25. https://doi.org/10.1016/j.neuron.2012.06.012, PMCID:PMC3743085, NIHMSID:NIHMS389499, PMID:22794257, 14 Aug 2013.

[4] K. McRae, M. Jones, and D. Reisberg (ed.), The Oxford Handbook of Cognitive Psychology. Oxford University Press. ISBN 9780195376746, 2013, pp. 206–216.

[5] P. Graf and D.L. Schacter, Implicit and explicit memory for new associations in normal and amnesic subjects, Journal of Experimental Psychology: Learning, Memory, and Cognition, 11(3), 501–518. https://doi.org/10.1037/0278-7393.11.3.501, PMID 3160813, 1985.

[6] H. Allain, A. Lieury, V. Thomas, J.M. Reymann, J.M. Gandon, and S. Belliard, Explicit and procedural memory in Parkinson's disease, Science Direct, Biomedicine & Pharmacotherapy (Elsevier), 49(4), 179–186. https://doi.org/10.1016/0753-3322(96)82618-7, 1995.

[7] D.L. Schacter, Implicit memory: history and current status, Journal of Experimental Psychology: Learning, Memory, and Cognition, 13(3), 501–518. https://doi.org/10.1037/0278-7393.13.3.501. S2CID 3728984. Archived from the original (.pdf) on 2009-02-19, 1987.

[8] M.T. Ullman, Contributions of memory circuits to language: The declarative/procedural model, Cognition, 92(1–2), 231–270. https://doi.org/10.1016/j.cognition.2003.10.008, PMID 15037131, S2CID 14611894, 2004.

[9] P. Graf, and G. Mandler, Activation makes words more accessible, but not necessarily more retrievable, Journal of Verbal Learning and Verbal Behavior, 23(5), 553–568. https://doi.org/10.1016/s0022-5371(84)90346-3, 1984.

[10] D.A. Sousa, Brain-compatible Activities, Grades 6–8, Book issue 6. ISBN: 978-1-63450-372-3, 2008.

[11] Pankaj Sah, The Brain Series, Issue Two, Chapter 2, How Memories Are Made. Queensland Brain Institute, Science of Learning Research Centre, The University of Queensland, Australia, downloaded in Sep 2023.

[12] J.J. Pillai and S.K. Mukherji, Functional Connectivity, Neuroimaging Clinics of North America, Vol. 27, n. 4. Elsevier, ISBN13: 978-0-323-54891-5, Nov 2017.

[13] J. Mill and J. Stuart Mill, Analysis of the Phenomena of the Human Mind. Good Press, ASIN: B081VNPLTH, 22 Nov 2019.

[14] I. Goodfellow, Y. Bengio, and A. Courville, Deep Learning Book. MIT Press, Chapter 2, 2016.

[15] F. Akram and J. Giordano, Research domain criteria as psychiatric nosology: Conceptual, practical and neuroethical implications, Cambridge Quarterly of Healthcare Ethics, 26(4), 592–601. https://doi.org/10.1017/S096318011700010X, 2017.

[16] www.uol.com.br/vivabem/noticias/deutsche-welle/2023/05/08/como-a-ciencia-explica-o-deja-vu-e-quando-ele-e-preocupante.htm, 2023.

[17] C.G. Jung and S. Freud, The Freud/Jung Letters: The Correspondence Between Sigmund Freud and C.G. Jung. The Princeton University Press, 1974.

[18] E. Bertelli, Anatomy of the Eye and Human Visual System. Eugenio Bertelli's Lab, Piccin, ISBN: 9788829929412, Sep 2019.

[19] S.A. Josselyn and P.W. Frankland, Memory allocation: Mechanisms and function, Annual Review of Neuroscience, 41(1). https://doi.org/10.1146/annurev-neuro-080 317-061956, Jul 2018.

[20] Y. Hu, K. Anders Ericsson, D. Yang, and C. Lu, Superior self-paced memorization of digits in spite of a normal digit span: The structure of a memorist's skill, Journal of Experimental Psychology: Learning, Memory, and Cognition , 35(6), 1426–1442. https://doi.org/10.1037/a0017395, PMID: 19857014, Nov 2009.

[21] D.A. Treffert, Accidental genius, AXS biomedical animation studio, SA Special Editions, 23(5s), 54–59. https://doi.org/10.1038/scientificamericangenius0115-54, 01 Jan 2015.

[22] N. Dellis and S. Stilwell, Memory Superpowers! An Adventurous Guide to Remembering What You Don't Want to Forget. Open Road Integrated Media, ISBN 9781683357766, 18 Aug 2020.

[23] F. Gobet, A. Snyder, T. Bossomaier, and M. Harré, Designing a "better" brain: Insights from experts and savants, Frontiers in Psychology, https://doi. org/10.3389/fpsyg.2014.00470, Source: PubMed, License CC BY 3.0,5:470, May 2014.

[24] J. Gallate, C. Wong, S. Ellwood, A. Snyder, and R.W. Roring, Creative people use nonconscious processes to their advantage, Creativity Research Journal (Taylor and Francis Group), 24(2–3), 146–151. https://doi.org/10.1080/10400419.2012.677282, Apr 2012.

[25] W. Langley, Is there a savant inside all of us? Financial Times, 04 Oct 2012.

[26] A. Hadhazy, Till where goes our memory capacity? [Até onde vai nossa capacidade de memória?], BBC Future, www.bbc.com/portuguese/noticias/2015/04/150408_ vert_fut_capacidade_cerebro_ml, 08 Apr 2015.

[27] S. Herculano-houze, The Human Advantage – A New Understanding of How Our Brain Became Remarkable. MIT Press, ISBN-13: 978-0262034258, 19 Apr 2016.

[28] J. Bahney and C.S. von Bartheld, The cellular composition and glia-neuron ratio in the spinal cord of a human and a nonhuman primate: Comparison with other species and brain regions, The Anatomical Record (National Institutes of Health [NIH]), 301(4), 697–710. https://doi.org/10.1002/ar.23728, 18 Nov 2017.

[29] F.A.C. Azevedo, et al., Equal numbers of neuronal and non-neuronal cells make the human brain and isometrically scaled-up primate brain, Journal of Comparative Neurology, 513, 532–541, 2009.

[30] J. Bahney and C.S. Bartheld, Validation of the isotropic fractionator: Comparison with unbiased stereology and DNA extraction for quantification of glial cells, Journal of Neuroscience Methods, 222, 165–174, 2014.

[31] S. Herculano-Houzel, The Human Advantage: A New Understanding of How Our Brain Became Remarkable. The MIT Press, 2016.

[32] E.R. Kandel, The Disordered Mind: What Unusual Brains Tell Us about Ourselves. FSG, 2018.

[33] J.E. Duque Parra, J.B. Ríos, and F.J.C. Peláez Cortes, and S.F. Ramón y Cajal, ¿Padre de la Neurociencia o Pionero de la Ciencia Neural? International Journal of Morphology, 29(4), 1202–1206. http://dx.doi.org/10.4067/S0717-95022011000400 022, ISSN 0717-9502 (On-line), 2011.

[34] W. Men, D. Falk, T. Sun, W. Chen, J. Li, D. Yin, L. Zang, and M. Fan, The corpus callosum of Albert Einstein's brain: Another clue to his high intelligence? Brain

(National Center for Biotechnology Information), 137(4), e268, https://doi.org/
10.1093/brain/awt252, Published online 2013 Sep 21, PMCID: PMC3959548,
PMID: 24065724, Apr 2014.

[35] E.R. Kandel and W.A. Spencer, Cellular neurophysiological approaches in the study
of learning, Physiological Reviews, 48(1), 65–134. https://doi.org/10.1152/phys
rev.1968.48.1.65, PMID: 4295027, Jan 1968.

[36] D. Andina, K. Fukushima, J.R. Peláez, and D.T. Pham, Neuroengineering: From
neurosciences to computations, International Journal of Neural Systems, 28(05,
Special Issue), 1803001, https://doi.org/10.1142/S0129065718030016, 2018.

[37] D.O. Hebb, The Organization of Behavior: A Neuropsychological Theory. John
Wiley and Sons, 1949.

[38] L.G. Reijmers, B.L. Perkins, N. Matsuo, and M. Mayford, Localization of a stable
neural correlate of associative memory, Science, 317(5842), 1230–1233. https://doi.
org/10.1126/science.11438, 31 Aug 2007.

[39] S. Fusi, P.J. Drew, and L.F. Abbott, Cascade models of synaptically stored memories,
Neuron (National Center for Biology of Medicine Information), 45(4), 599–611.
https://doi.org/10.1016/j.neuron.2005.02.001, PMID: 15721245, 17 Feb 2005.

[40] I.T. Bayazitov, R.J. Richardson, R.G. Fricke, and S.S. Zakharenko, Slow presynaptic
and fast postsynaptic components of compound long-term potentiation, Journal
Neuroscience, 27, 11510–11521, 2007.

[41] E.S. Boyden, A. Katoh, J.L. Pyle, T.A. Chatila, R.W. Tsien, and J.L. Raymond,
Selective engagement of plasticity mechanisms for motor memory storage, Neuron,
51, 823–834, 2006.

[42] C.J. Magnus, P.H. Lee, D. Atasoy, H.H. Su, L.L. Looger, and S.M. Sternson, Chemical
and genetic engineering of selective ion channel-ligand interactions, Science,
333, 1292–1296. https://doi.org/10.1126/science.1206606, PMID: 21885782,
PMCID: PMC3210548, 2011.

[43] Q. Bu, A. Wang, H. Hamzah, A. Waldman, K. Jiang, Q. Dong, R. Li, J. Kim, D.
Turner, and Q. Chang, CREB signaling is involved in Rett syndrome pathogenesis,
Journal of Neuroscience, 37(13), 3671–3685. https://doi.org/10.1523/JNEURO
SCI.3735-16.2017, PMCID: PMC5373141, PMID: 28270572, 29 Mar 2017.

[44] T.J. Shors and L.D. Matzel, Long-term potentiation: What's learning got to do with
it? Behavioral Brain Science, 20(4), 597–614; discussion 614–655. https://doi.org/
10.1017/s0140525x97001593, PMID: 10097007, Dec 1997.

[45] D. Purves, G.J. Augustine, D. Fitzpatrick, L.C. Katz, A.-S. LaMantia, J.O. McNamara,
and S.M. Williams, Neuroscience, 2nd ed. Sinauer Associates, ISBN-10: 0-87893-
742-0, 2001.

[46] B. Christen, F.F. Damberger, D.R. Pérez, and K. Wüthrich, Structural plasticity of the
cellular prion protein and implications in health and disease, PNAS, 110 (21) 8549–
8554, https://doi.org/10.1073/pnas.1306178110, Contributed by Kurt Wüthrich, 06
May 2013.

[47] A.J. Silva, How one memory attaches to another. In revolutions in science: Discoveries
that could change everything, Scientific American, 27(3s), July 2018.

[48] D.J. Cai, D. Aharoni, T. Shuman, J. Shobe, J. Biane, W. Song, B. Wei, M. Veshkini,
M. La-Vu, J. Lou, S. Flores, I. Kim, Y. Sano, M. Zhou, K. Baumgaertel, A. Lavi,
M. Kamata, M. Tuszynski, M. Mayford, P. Golshani, and A.J. Silva, A shared
neural ensemble links distinct contextual memories encoded close in time, Nature,
534(7605), 115–118, 23 May 2016.

[49] T.P. Reber, J. Faber, J. Niediek, J. Boström, Christian E. Elger, and Florian
Mormann, Single-neuron correlates of conscious perception in the human medial

temporal lobe, Current Biology, 27(19), 2991–2998, https://doi.org/10.1016/j.cub.2017.08.025, 2017.

[50] B. Das, Framework for Conscious Information Processing. Cornell University, arXiv:q-bio/0702033v2, https://doi.org/10.48550/arXiv.q-bio/0702033, www.researchgate.net/publication/2183806, 18 Oct 2015.

[51] D.J. Chalmers, The representational character of experience, in Brian Leiter (ed.), The Future for Philosophy. Oxford University Press, 2004.

5 Principles of Natural Intelligence

5.1 INTRODUCTION

The scientific study of the biological basis of natural intelligence has contributed for decades to the understanding of individual differences in the cognitive abilities of each human being. In particular, the continued development of electrophysiological neuroimaging and genetic methods has created new opportunities to gain insights into the details of pressing questions. The result of this may allow this field to move closer to a comprehensive theory that explains how genotypes influence human intelligence through intermediate biological and cognitive processes, the so-called endophenotypes. Some studies already summarize reference findings for electro-physiological neuroimaging and genetic research. There are, however, four issues that must be clarified in order to better implement the results of these studies: 1) the robustness of the research results; 2) the relationship between neural parameters and cognitive processes; 3) the promising methodological developments; and 4) the development of a generic theory [1]. Therefore, it is necessary to seek a better inter-pretation for such brain actions within the brain storage of ideas and human impulsive knowledge, which is the main approach of this chapter.

5.2 INFORMATION FROM HUMAN AND ANIMAL BRAIN SENSORS

According to Howard Gardner [2], natural intelligence is the ability of human beings to identify, classify and manipulate elements of the environment, objects, animals or plants. In accordance with this and asked about the source of his genius, Albert Einstein (1879–1955) responded bluntly: "I believe in intuitions and inspirations. Sometimes I feel that I am right. I don't know if I am," he told the US magazine *Saturday Evening Post* in 1929. The physicist was far from being the only one to adopt this philosophy and stated that it was much easier to trust these intuitions and check them later than to dismiss them prematurely.

As discussed in Chapter 1, the five well-known senses of living beings are touch, hearing, sight, smell and taste. These senses seem to depend on vibrations that are nothing more than repetitive frequencies. Each sense is subject to a range of

DOI: 10.1201/9781003604037-5

characteristic frequencies. The frequencies of touch are the lowest ones and depend on the vibration of the things that are touched, ranging from 0 Hz to around 200 Hz. This can include the temperature differences between the organs and the texture of the place touched. Hearing or sound frequencies range from 15 Hz to somewhere around 25 kHz. The universally adopted standard ranges from typically 20 Hz to 20 kHz, which appears to be the perceptible range of frequencies for most people. Vision frequencies are typically a combination of the frequencies of the basic colors: red, green and blue (RGB). These frequencies typically range from 405 THz (1 THz = 10^{12} Hz) (red color) to 790 THz (violet color). Beyond these frequencies, few people can perceive any coloration. The frequencies of smell and taste need to be better studied. For extrasensory perception, probably other frequencies could explain this so-called 'sixth sense', which often refers to what is known as spirituality. It is often said that women have a stronger sixth sense than men [3–5].

The brain is in constant activity and neurons are always communicating and exchanging information through brain waves by electrochemical signals. Some regions will be working with more or less vigor, but none are completely 'not working'. These waves are classified into five types according to their effects on human behavior: alpha, beta, gamma, delta and theta.

Alpha-type waves are intermediate frequencies, with oscillations that vary between 8 Hz and 13 Hz and are more related to relaxation. They help the body and mind to achieve states of bodily and mental relaxation in stressful situations. Even so, they require brain activity that can be intense, like what people use in meditation, prayer, using intuition or pursuing a creative hobby.

Beta-type waves are a little broader and have an intensity of between 12 Hz and 33 Hz. Even so, they are considered high to moderate frequency waves and are very focused on studies and knowledge as well. They are very suitable for carrying out cognitive tasks that require specific skills, such as giving a presentation and taking a proficiency test, among other examples.

Gamma waves are the fastest and have a higher electrical frequency, reaching 100 Hz, while the other waves vary between 3 Hz and 20 Hz. Therefore, they are perfect waves for carrying out activities that require high concentration, for in-depth studies and important decision-making. Because its information processing is very fast, the gamma wave is the one that most helps the brain when it needs to carry out activities that require reasoning.

Delta waves are those with the lowest frequency, from 1–3 Hz. Therefore, they help human beings to enter a state of total and deep relaxation, both mentally and bodily. These are the waves that manifest themselves in deep sleep, always taking care of vital activities, such as breathing, heartbeat and blood pressure.

Theta waves have frequencies below those of alpha waves, ranging between 3.5 Hz and 8 Hz. They also help people achieve states of relaxation but, unlike alpha waves, they command body relaxation at the end of the day. These waves refer to relaxation after moments of many activities and are essential for sleep and rest.

It is proven that people who suffer from a deficiency related to the sensory system have a form of compensation and end up developing more of their other senses. For example, a blind person better develops their ability to hear or even feel books in

Braille. This form of writing is done from right to left, through the points created with a punch. To read, the blind person turns the page over and feels the relief formed on the back.

Like the ears and eyes, the intestine and skin also have several organic similarities between them, such as, they are very rich in vessels and nerves; they are contact organs, that is, they are responsible for one of the communication forms between humans and the environment and are filled with a large number of microorganisms. The diversity of microorganisms found on the skin is similar to that in the intestine. Recent studies suggest that the entire genome of microorganisms that make up the skin (microbiome) can be influenced by the intestine with the possibility of altering human skin [6,7]. The intestinal microbiota is made up of a population of microorganisms, mainly bacteria, that live in the human intestines, playing a fundamental role both in the intestine and in its distance and sensitivity from other organs, including the skin [6,7].

The neurophysiological trace of brain activity after a cardiac arrest and during a near-death experience (NDE) is not well understood. Although a hypoactive state of brain activity has been assumed, experimental animal studies have shown an increase in activity following a cardiac arrest. This occurs particularly in the gamma band, resulting from hypercapnia before and after the interruption of cerebral blood flow shortly after cardiac arrest. It appears that no detailed studies have yet investigated this issue in humans. There exist continuous electroencephalographic (EEG) recordings of a dying human brain obtained from an 87-year-old patient who went into cardiac arrest following a traumatic subdural hematoma. Although this study is in its initial stages worldwide for measuring living brain activity during the death process in humans, it is already possible to probe similar changes in gamma oscillations that were previously observed in rats kept in controlled environments. This means that it is possible that, at death, the brain organizes and executes a biological response that can be conserved across species. These measurements however, are based on a single case and come from the brain of a patient who suffered injuries, seizures and swelling, which further complicates reliable interpretation of the results. Ajmal Zemmar, one of the pioneers in these studies, plans to investigate more cases on this subject and sees these results as a source of hope for elucidating how the brain actually disappears after physical death. He states that one thing that can be learned from this is that although loved ones may have their eyes closed and be ready to die, their brains may be replaying some of the best moments they had in their lives [8,9].

Close to death, human beings experience an increase in absolute power in gamma activity in the narrow and broad bands and a decrease in theta power after the suppression of bilateral hemispheric responses. After cardiac arrest, the power types alpha, beta, gamma and delta decrease, but a higher percentage of relative gamma power is observed when compared to the interictal interval. Cross-frequency coupling revealed a modulation of gamma activity in the left hemisphere by alpha and theta rhythms in all windows, even after cessation of cerebral blood flow. The strongest coupling was observed for narrow- and broad-band gamma activity by alpha waves during left-sided suppression and after cardiac arrest. Despite the influence of neuronal injury and swelling, the data now provided the first evidence of what happens in a dying human brain in a real-life non-experimental intensive care clinical setting. These results allow scientists to defend the idea that the human brain may have the capacity to generate well-coordinated activity during the period of near death [10,11].

It is known that neurons communicate through electrical impulses produced by ion channels that control the flow of ions such as potassium and sodium, this being the chemical side of this relationship. In a surprising new discovery, MIT neuroscientists have shown that human neurons have many fewer of these channels than expected compared to neurons in other mammals. The researcher hypothesizes that this reduction in channel density may have helped the human brain evolve to function more efficiently, allowing it to divert resources to other energy-intensive processes needed to perform some more complex cognitive tasks. "If the brain can save energy by reducing the density of ion channels, it can expend that energy on other neuronal or circuit processes", says Mark Harnett, associate professor of brain and cognitive sciences, a fellow at MIT's McGovern Institute for Brain Research and senior author of this study [12].

Harnett and his colleagues analyzed neurons from ten different mammals. This was the most extensive electrophysiological study of its kind to date, and they identified a 'construction plan' applicable to every species they observed except humans. They found that as the size of neurons increases, the density of channels found in the neurons also increases. However, human neurons have proven to be a notable exception to this rule. Neurons in the mammalian brain can receive electrical signals from thousands of other cells, and this input determines whether, if necessary, they fire an electrical impulse called an action potential. In 2018, Harnett and Beaulieu-Laroche discovered that human and rat neurons differ in some of their electrical properties, particularly in parts of the neuron called dendrites that are tree-like antennas that receive and process information from other cells.

The increase in channel density across species is surprising, Harnett says, because the more channels there are, the more energy is needed to pump ions in and out of the cell. However, it started to make sense when researchers started thinking about the number of channels in the overall cortex volume. In the tiny brain of the Etruscan shrew, packed with very small neurons, there are more neurons in a given volume of tissue than in the same volume of tissue in the rabbit brain, which has much larger neurons. Nevertheless, because rabbit neurons have a higher density of ion channels, the density of channels in a given volume of tissue is the same in both species or in any of the non-human species so far analyzed by researchers. This construction plan is consistent across nine different species of mammals and it appears as if the cortex is trying to maintain the same number of ion channels per unit volume in all species. This means that, for a given volume of cortex, the energetic cost is the same, at least for the ion channels.

The human brain represents a marked deviation from the construction plan relative to all animals. In humans, ion channels have the functions of conducting, recognizing and selecting specific ions for communication within the brain. Opening or closing occurs in response to electrical, mechanical and/or chemical stimuli, acting in the transmission of energy through the membranes of various tissues. Interestingly, instead of a progressive increase in the density of these ion channels, the researchers found that there is a dramatic decrease in the expected density of them for a given volume of brain tissue. It is thought that this lower density may have evolved as a way to expend less energy on pumping ions to allow the brain to use that energy for other functions, such as creating some more complicated synaptic networks between neurons or firing action potentials at a higher rate to become more energy efficient, using less adenosine triphosphate (ATP) per volume compared to other species. This molecule constitutes the main form of chemical energy, since its hydrolysis is highly exergonic. That is, when undergoing the fission process caused by water (hydrolysis), this molecule releases large amounts of free energy.

5.3   MANAGEMENT OF BRAIN INFORMATION

Wisdom is the quality of having extensive and deep knowledge of various things or of a particular topic without needing to have a direct relationship with intelligence. Intelligence is the ability to reason, understand and resolve new problems and conflicts, as well as being able to adapt with certain ease to new situations. It cannot be said that a computer is intelligent, but it can be very wise because it has received a lot of information and knowledge. In other words, intelligence results from the functioning of the brain structure, while wisdom is just the amount of data stored in this structure. Evidently, children with primarily naturalistic intelligence show an inclination toward the natural world as well as toward things that humans have already created. They go beyond superficial observations and go deeper to know about things working and their nature. They also have a tendency to classify objects and classify them into categories according to their still immature conceptions.

In relation to the animal kingdom, but also including human beings and plants, children with expressive natural intelligence express their desires by observing nature, enjoying having pets and being fascinated by the growth of plants. They like to explore and discover natural environments and other forms of life because they are coming into contact with a whole new world, and it is common to find them carrying out improvised experiments with their toys and pets. Children like these have a lot of fun observing details of how cats feed, trying to imitate dogs, watching birds fly and how cows can give milk, examining the size of elephants and crocodiles, or simply playing with insects.

Regarding children's education, there are different activities that can help them develop their natural intelligence, which include [13,14]:

1) Visits to tourist places, museums, amusement parks, public works and others;
2) Generic observation as a way to strengthen this type of intelligence by communicating closely with the surrounding environment;
3) Hobbies such as planting flowers, vegetable gardens, plant decorations at home or collecting fossils, stones, leaves, etc.;
4) Getting closer to nature through walks in the woods or mountains, camping, scouting, visits to zoos and aquariums, botanical gardens, etc.;
5) Classification of what exists around them based on cell phones, books or the internet, encouraging them to record what they learn and create new versions of it all;

6) Facilitation of observation means by making materials such as magnifying glasses, microscopes, binoculars and small home laboratories available in such a way as to allow common experiences with the scientific and techno-logical world.

An interesting aspect of the accumulation of information in the brain is the ability to memorize what the intestines, ears, eyes and other human senses record. There may be a certain saturation and selection in the amount of data or information stored that may cause difficulties in immediate access. This is because from birth, or even during pregnancy, people receive and accumulate information from the environment and the mother's womb through the senses. All of this is stored in the brain, whose capacity is limited to approximately 100 Tbytes (1 Tbyte = 10^{12} bytes) of neurons. (Note that some IT companies are already announcing the production of solid-state drives (SSDs) with a capacity of 100 Tbytes or much more). Of all this, only something like 10% can be managed consciously.

With the accumulation of information in a single human memory, sometimes, spontaneously, the brain, or the 'manager' of the information location requested from the brain, has a certain difficulty in finding an immediate solution in some unusual situation. This is due to the natural quantitative limitation of the brain's 'manage-ment' to quickly search for information and thus make it readily available. It may happen that the search time is short because this random search was initiated occa-sionally by the desired information or by the most recent information or even because of the small amount of information available. Otherwise, it may take an appreciable amount of time. This amount of information may explain why elderly people cannot immediately remember some of the things they are asked or what they want to say at a given moment. After some time, these memories emerge unexpectedly for them, often when the subject is no longer of interest in the conversation.

Some of the historical figures considered to have the highest naturalistic intel-ligence include Charles Darwin (1809–1882), the famous English naturalist who developed the theory of evolution through natural selection in his work *The Origin of Species* (1859), or Alexander von Humboldt (1769–1859), the Prussian explorer and naturalist considered to be the father of modern geography. Important figures in more recent times include the French oceanographer Jacques Cousteau (1910–1997) and the American astronomer Carl Sagan (1934–1996), great communicators who helped awaken and enhance the naturalistic intelligence of an entire generation, and Stephen William Hawking (1942–2018), internationally recognized for his contribu-tion to science, being one of the most renowned scientists of the current century. With so much knowledge acquired throughout life, it is clear that this amount of informa-tion accumulated from a young age often makes it difficult for the memory manager to find it immediately among so much other information [15].

Of note in the management of brain information are those impulsive actions to which every human being is subject with greater or lesser intensity. Such unusual impulses are the result of the accumulation of information in the memory. These impulses often make the person think that it was an 'angel' who previously informed them about a situation or something not deliberately planned. For example, there might be someone who does not want to arrive somewhere because of the abundant unconscious information previously accumulated in the brain telling them that there will be a complicated problem to be faced when getting to that place. It may happen that, for example, when driving home by car, the driver will face some unwanted problems. This manifestation of signals can happen unconsciously at the car speed, causing it to be manifested by a traffic light suddenly closing, a puppy unexpectedly crossing the street, an intersection being crowded with vehicles, the guard blowing his whistle at someone and so on. All of this can be unconsciously arranged by the driver's own brain information according to the content of his or her thoughts, a driver who does not want to face the problem that awaits him or her on arrival at the place being traveled to. If the angel is contradicted, there will most likely be problems. A situation contrary to this can also happen. That is, for the best solution to a certain problem in the brain, the living person will only need to go to a certain place or make a certain decision, defined by subconscious information, which is manifested by facilities on the way or something this way, and the problem will be resolved satisfactorily.

5.4 INTUITION OR UNCONSCIOUS OR SUBCONSCIOUS REASONING

According to the Merriam-Webster dictionary, intuition is the mental power or faculty of attaining direct knowledge or cognition without evident rational thought and inference, or an immediate apprehension or cognition, or knowledge of a conviction obtained by intuition. However, because it is so abstract, different fields use the word 'intuition' in very different ways, including, but not limited to, direct access to unconscious knowledge, unconscious cognition, gut feelings, inner sensation, internal insight for pattern recognition unconscious, experience and also the ability to understand something instinctively, without any need for conscious reasoning. Intuitive knowledge tends to be close to reality, but rarely accurate [16–20].

Intuitions are often invoked independently of any particular theory of how they provide evidence for claims. Note trhat there are many divergent accounts of what types of mental state intuitions are, ranging from mere spontaneous judgment to special presentation of a strictly necessary truth. In recent years, several philosophers, such as George Bealer [21], have attempted to defend appeals to intuition against Willard Van Orman Quine [22], who developed what are known as Quinean doubts about conceptual analysis. A different challenge to appeals to intuition has come more recently from experimental philosophers, who argue that appeals to intuition can be better informed by the methods of social science [23,24].

To explain someone's mental capacity, it is necessary to take into account that the human brain is loaded from birth, and even before, with an enormous amount of information coming from the most varied sources from inside the mother's womb. In other words, the creative environment has much to do with each person's mental capacity. Examples include pianists from a young age, child mathematicians, designers practically from birth, prominent singers from an early age and many others, like some of the famous cases presented in Chapter 10 of this book. Among the origins of learning are caloric, sound, visual, sensitive, vibrational, educational, emotional and gustatory sources, which add to each person's imagination to a greater or lesser extent depending on their experience and opportunities.

Evidence shows self-similar patterns of conductive resonances repeating across the terahertz, gigahertz, megahertz, kilohertz and hertz frequency ranges in microtubules. These conductive resonances apparently originate from quantum dipolar oscillations in terahertz and optical interactions between Pi electron resonance clouds of the aromatic amino acid rings of tryptophan, phenylalanine and tyrosine within each tubulin, which is the component subunit of microtubules, and the most abundant protein of the brain. Evidence from cultured neuronal networks also now shows that gigahertz and megahertz oscillations in dendritic–somatic microtubules regulate specific firing of distal axonal branches, causally modulating synaptic and membrane activities. Therefore, the brain should be viewed as a scale-invariant hierarchy, with quantum and classical processes critical for rational consciousness and cognition originating in microtubules within neurons [25,26].

Good management of brain information is something that can be guided by human beings themselves according to their culture. To achieve this, cultivating positive ideas can help charge the subconscious with positive solutions to provide a good answer to life's daily problems and vice versa. Going along this line, repetitive prayers, such as those preached by religions, seem to be an ancient wisdom for convincing the brain that with good ideas someone can have good solutions and become a positive help. Perhaps the people who preached repetitive prayers since ancient times did not initially understand the reason for this cerebral need to preach positive ideas, but they just repeated them without criticism. Perhaps it was a poeticized form of everything they learned over time and, probably, it is still happening worldwide today [27,28]. Within this line of reasoning, the Oxford English Dictionary gives an important definition of what 'culture' is: the 'cultivation' or development of the mind, faculties, manners, etc., for improvement or refinement by education and training, development and refinement of mind, tastes and manners. It follows that people become educated through training in a variety of activities, such as mathematics, medicine, clothing, customs, arts, ways of interacting with people, use of technology, relationships, computing and learning shared ideas, beliefs, philosophies and religions [27,28].

The conscious mind is responsible for rationalization and logical thinking, while the subconscious mind is responsible for involuntary actions. Most spiritually developed people have identified the basis of these two terms and have enhanced the potential to establish a link between them by continuous mental training methods, which are recognized as ways to improve the quality of life. Therefore, the difference between the conscious and subconscious mind is that the conscious mind is responsible for the awareness of an incident in the present that includes some of the internal mental functions in addition to the external events. For example, in a conscious mind, a person can focus on writing a letter while enjoying the window reflecting its rays onto the table. In turn, the subconscious is not aware of both internal mental functions and external events, with it appearing as if there is no connection between them, but being just a storehouse of information.

In psychology, the terms 'conscious' and 'subconscious' are used variably to define state of mind. Due to several similar, complex and unclear features they share, it can be highly confusing to identify each one, even for experts in the field. However, it seems to be common sense, as far as the theoretical aspect is considered, that the human mind is divided into three states known as conscious mind, subconscious mind and unconscious mind. Several research studies have been carried out to differentiate and analyze these terms individually with regard to how they define human attitudes and behavioral patterns (see Chapter 7). The biggest difference between the conscious and subconscious minds appears to lie mainly in basic human functions and mental processes.

In summary, unconscious reasoning occurs in at least three different ways: 1) it can be unconscious as a whole and influence behavior outside of consciousness; 2) it can produce intuitions of which the person is aware, although they are not aware of inferring them at a given moment; and 3) it can support even the simplest of human conscious inferences. The conclusion is that mental architecture theory states that inferences can be initiated consciously or unconsciously [29].

It is interesting to mention here something about Dr Joseph Murphy (1898–1981), from the University of India, a writer, professor and lecturer who became known for his bestsellers of high spiritual value. This author was one of the most profound experts on mental and spiritual laws and remains to this day a source of inspiration for a legion of followers. His courses on the power of the subconscious have always attracted many people, and his work continues to win thousands of readers around the world.

According to Murphy, any question arises from how many times has a person asked the Universe for something and never received an answer. Why do some people seem to get everything they want, while others remain sad and miserable? Why do some people receive a cure for a serious illness, while others die from something relatively simple? Murphy [30,31] stated that the secret is within each person, more specifically in the subconscious. He adds, "That the mind is more powerful than we imagine, as it is with it that we can have happiness, peace of mind, riches, a cure for illness, misfortune and much more" [30]. The author himself experienced a cure for sarcoma through the power of his subconscious and from then on, he allowed his mind to govern his vital functions. Joseph Murphy wrote many books, 286 in total, addressing this very special topic so that the desired objectives could be achieved [30,31].

It is important to include at this point the instincts of living beings as natural intelligence. An instinct is a characteristic of each living organism to perform a certain complex behavior containing innate elements. It is a sequence of short or medium-sized actions carried out in response to a stimulus clearly defined by such species. One of these sequences is known as fixed action pattern (FAP), which is characterized by similar actions in the face of similar external phenomena.

A behavior is instinctive if there is no previous experience that can be characterized by innate biological factors. An interesting example of this are sea turtles that instinctively move towards the ocean from a young age. Recently born birds waiting for their food in their nests. A marsupial climbing into its mother's pouch at birth. Animals' aggressive instincts include self-defense, fighting among themselves, typical dominance behavior, construction of their habitats and internal escapes. These performances can be altered by experiences, becoming more or less sharpened by certain characteristics common to their species. Ways of retreating to sleep, escaping scenes that could be dangerous, keeping warm in the cold or reacting in the heat are some of these volitional characteristics [32,33].

The instinct is formed by previous knowledge acquired since the mother's womb, and even before that, by knowledges accumulated by the mother's brain. This knowledge depends on each species based on their size and sex, the environment where they came from, their body structure, their natural bodily fluids and many other body/brain characteristics.

Norman Doidge, a research psychiatrist and psychoanalyst, describes in detail in a book the discovery that the human brain can change itself, as reported in stories of scientists, doctors and patients who, together, carried out this type of transformation [34]. Without surgery or medication, several researchers used a previously little-known ability to change the brain. Some were patients who appeared to have incurable brain problems and others were people without specific problems. The latter just wanted to improve the performance of their brains or, better yet, preserve them as they aged. Until then, knowledge on this subject was that, after childhood, the brain only changed when it began to decline; that is, when the brain cells began to fail in their development, or if they were injured or dead, and could not be replaced. It was imagined that the brain could also not change its structure with its use and find a new way to compensate for some part that was damaged, with the assertion that people who were born with brain or mental limitations, or who suffered brain damage, would be thus limited or damaged for life. It was thought that the brain could not be improved or preserved through mental activities or exercises. The belief that the brain could not change had three main sources: the type of brain-damaged patients who rarely made a full recovery, the inability to observe the microscopic activities of the living brain and the idea that the brain functioned like a digital machine subject to defects and could not change or grow on its own.

As Doidge describes it, patients who do not progress psychologically as much as expected find that their problems are deeply 'hardwired' in an unchanging brain. With the evolution of artificial intelligence, hardwiring was thought of as a metaphor that the brain is like the hardware of a computer and, therefore, has a specific and immutable composition [34]. To allow a parallel on this subject, Chapter 6 discusses in more detail how artificial intelligence works.

Doidge also describes an experiment carried out by Drs Guang Yue and Kelly Cole regarding how to use muscles to strengthen them by analyzing two groups, one that did physical exercises and the other that imagined just doing exercises. Both groups exercised a finger muscle from Monday to Friday for four weeks. The physical group attempted 15 maximal contractions, with 20 seconds of rest between each one. The mental group simply imagined doing 15 maximal contractions, with a 20 second rest between each one, while also imagining a voice shouting at them, "Harder! Harder! Harder!" At the end of the study, the subjects who exercised increased their muscle strength by 30%, as expected. Those who only imagined doing the exercise in the same period increased their muscle strength by 22%. The explanation lies in the brain's motor neurons that 'program' movements. During these imaginary contractions, the neurons responsible for chaining sequences of instructions for movements are activated and strengthened, resulting in an increase in strength when the muscles are contracted. This research allowed the development of the first machines that actually 'read' people's thoughts. Thought translation machines access motor programs in a person or animal imagining an act, decoding the distinct electrical signature of the thought, and transmitting an electrical command to a device that puts the thought into action. These machines work this way because the brain is plastic and physically changes its state and structure as the person imagines in such a way that it can be recorded with electronic measurements [34].

If the perception is that the brain is really a very complex organ, the mind is even more so, as expressed in the words of Nicolelis (2011), that at the same time as the brain is the human essence, the behaviors and individual characteristics of each one, it also makes them part of the collective. By this, it was meant that human beings are unique and special, the results not only of their genes, but also of situations such as their opportunities, impressions, experiences, social relationships, pain, violence and learning [33,34]. Despite being unique, human beings will never be finished and ready, quite the contrary, they will always be subject to transformations, evolutions and innovations. They will always be aware of new solutions to life's problems and will seek to evolve to have better conditions for brain and mental functions, thus having greater longevity, health and quality of life.

REFERENCES

[1]   K. Hilger, F.M. Spinath, S. Troche, A. Schubert, The biological basis of intelligence: Benchmark findings, Science Direct, Book Intelligence (Elsevier), 93, 101665, https://doi.org/10.1016/j.intell.2022.101665, Jul–Aug 2022.

[2]   H.E. Gardner, Frames of Mind: The Theory of Multiple Intelligences, 3rd ed. Basic Books, ASIN: B01K3NO9P0, ISBN-13: 978-0465024339, 29 Mar 2011.

[3]   National Aeronautics and Space Administration, Science Mission Directorate. (2010). Infrared Waves. NASA Science website: http://science.nasa.gov/ems/07_in fraredwaves, retrieved 14 Jan 2025.

[4]   H.H. Skilling, Fundamentals of Electric Waves, 2nd Edition, book, John Wiley and Sons, London, 1948.

[5]   J. Allen, Ultraviolet Radiation: How It Affects Life on Earth. Earth Observatory, NASA, 06 Sep 2001.

[6]   M. Maguire and G. Maguire, The role of microbiota, and probiotics and prebiotics in skin health, Archives of Dermatological Research, 309(6), 411–421, 2017.

[7]   B. Das and G.B. Nair, Homeostasis and dysbiosis of the gut microbiome in health and disease, Bioscience Journal, Uberlância-MG-Br, 44(5), 2019.

[8]   R. Vicente, M. Rizzuto, C. Sarica, K. Yamamoto, M. Sadr, T. Khajuria, M. Fatehi, F. Moien-Afshari, C.S. Haw, R.R. Llinas, A.M. Lozano, J.S. Neimat, and A. Zemmar, Enhanced interplay of neuronal coherence and coupling in the dying human brain, Frontiers in Aging Neuroscience, 14, Article 813531, https://doi.org/10.3389/fnagi.2022.813531, www.frontiersin.org, Feb 2022.

[9]   D. Mobbs and C. Watt, There is nothing paranormal about near-death experiences: how neuroscience can explain seeing bright lights, meeting the dead, or being convinced you are one of them, Trends in Cognitive Sciences, 15, 447–449. https://doi.org/10.1016/j.tics.2011.07.010, 2011.

[10]   Y. Zhang, Z. Li, J. Zhang, Z. Zhao, H. Zhang, M. Vreugdenhil, et al., Near-death high-frequency hyper-synchronization in the rat hippocampus, Frontiers in Neuroscience, 13, 800. https://doi.org/10.3389/fnins.2019.00800, 2019.

[11]   E. Facco and C. Agrillo, Near-death experiences between science and prejudice, Frontiers in Human Neuroscience, 6, 209. https://doi.org/10.3389/fnhum.2012.00209, 2012.

[12]   L. Beaulieu-Laroche, E.H.S. Toloza, M. Van der Goes, M. Lafourcade, D. Barnagian, Z.M. Williams, E.N. Eskandar, M.P. Frosch, S.S. Cash and M.T. Harnett, Enhanced dendritic compartmentalization in human cortical neurons, Journal Cell (Elsevier), 175(3), 643–651. e14. https://doi.org/10.1016/j.cell.2018.08.045, 18 Oct 2018.

[13]   M.C. Hall, Multiple Intelligences: Teaching Kids the Way They Learn. Frank Schaffer Publications, 1999.

[14]   S. Suazo-Díaz, Inteligencias múltiples: manual práctico para el nivel elemental. La Editorial, Universidad de Puerto Rico, 2006.

[15]   A. Sponsel, Darwin and Humboldt, in Alison M. Pearn (ed.), A Voyage Round the World. Cambridge University Press, 2009, pp. 13–15.

[16]   S. Epstein, Demystifying intuition: What it is, what it does, and how it does it, Psychological Inquiry, 21(4), 295–312. https://doi.org/10.1080/1047840X.2010.523 875, S2CID 145683932, 30 Nov 2010.

[17]   S. Aurobindo, The Synthesis of Yoga. Sri Aurobindo Ashram Trust. ISBN 978-0-9415-2465-0. Retrieved 26 Dec 2014, 1992, pp. 479–480.

[18]   A.D. Rosenblatt and J.T. Thickstun, Intuition and consciousness, Psychoanalytic Quarterly, 63(4), 696–714, 1994.

[19]   G.J. Gigerenzer, Gut Feelings: The Intelligence of the Unconscious. Penguin. ISBN 978-0-670-03863-3, Oct 2007.

[20]   N. Angier, Intuition and math: A powerful correlation, The New York Times. Retrieved 2022-09-27, ISSN 0362-4331, 16 Sep 2008.

[21]   R.C. Koons and G. Bealer (eds.), The Self-consciousness Argument: Functionalism and the Corruption of Intentional Content, The Waning of Materialism: New Essays. Oxford University Press, 2010.

[22]   W.V.O. Quine, On What There Is, Review of Metaphysics, 2(5), 21–38, 1948.

[23]   M. Lynch, Trusting intuitions, in P. Greenough and M. Lynch (ed.), Truth and Realism. Oxford University Press, 2006, pp. 227–238.

[24]   S. Loev, Affectivism about intuitions, Synthese, 200, 274. https://doi.org/10.1007/s11229-022-03749-0, 2022.

[25]   S. Hameroff, Consciousness, cognition and the neuronal cytoskeleton – A new paradigm needed in neuroscience, Frontiers in Molecular Neuroscience, 15 https://doi.org/10.3389/fnmol.2022.869935, 2022.

[26]   M.C. da Silva, Biotecnologia, Blog do Profissão Biotec, ISSN 2675-6013, Blog do Profissão Biotec, Neurociência, 9, 2022.

[27]   R.V. Gulick, Consciousness and cognition, in The Oxford Handbook of Philosophy of Cognitive Science, https://doi.org/10.1093/oxfordhb/9780195309799.013.0002, 01 May 2012, pp. 19–40.

[28]   K. Grill-Spector, R. Henson and A. Martin, Repetition and the brain: neural models of stimulus-specific effects, Trends in Cognitive Sciences, 10(1), 14–23, 1364-6613/$, Elsevier, https://doi.org/10.1016/j.tics.2005.11.006, www.sciencedirect.com, Jun 2005.

[29]   P.N. Johnson-Laird, Chapter 5, Intuitions and unconscious reasoning, in Book How We Reason, https://doi.org/10.1093/acprof:oso/9780199551330.003.0005, Oct 2008, pp. 60–72.

[30] J. Murphy, O Poder do Subconsciente, best seller; 111ª ISBN-10: 8546501459, ISBN-13: 978-8546501458, 22 Jan 2019.

[31] J. Murphy, Liberte o Poder do seu Subconsciente, ISBN-10: 8576847647, ISBN-13: 978-8576847649, 16 Feb 2017.

[32] M.H. Bornstein, D.L. Putnick, P. Rigo, G. Esposito, J.E. Swain, J.T.D. Suwalsky, X. Su, X. Du, K. Zhang, L.R. Cote, N. de Pisapia, and P. Venuti, Neurobiology of culturally common maternal responses to infant cry, Proceedings of the National Academy of Sciences, 114(45), E9465–E9473, 2017.

[33] T. Li, M. Horta, J.S. Mascaro, K. Bijanki, L.H. Arnal, M. Adams, R.G. Barr and J.K. Rilling, Explaining individual variation in paternal brain responses to infant cries, Physiology & Behavior, 193(A), 43–54, https://doi.org/10.1016/j.phys beh.2017.12.033, PMC: 6015531, PMID: 29730041, 2018.

[34] N. Doidge, The Brain that Changes Itself: Stories of Personal Triumph from the Frontiers of Brain Science. Penguin Publishing Group, ISBN-13: 9781101147115, 18 Dec 2007.

[35] Nicolelis, M., Muito além do nosso eu. edition Schwarcz Ltda, Companhia das Letras, ISBN: 978-85-359-1873-1, 2011.

6 Principles of Artificial Intelligence

6.1 INTRODUCTION

The branch of science & technology that has been dedicated to creating machines that learn and think intelligently is known as artificial intelligence (AI). AI is an attempt to make a computer, robot or other piece of technology capable of 'making decisions', that is, processing data in the same way as humans. Many believe that such 'decision-making' has the same complexity as 'thinking' and that it is still a way of replacing slavery regimes so that others can do the work of the rulers. Therefore, AI has to study how the human brain 'thinks', learns, performs tasks and makes decisions based on accumulated and available data when trying to solve problems or perform a task that was once a primordial attribute of human beings. Therefore, the objective of AI can be established as existing to improve technology by adding functionalities related to human acts of reasoning, learning and problem solving. In this chapter, some practical examples are discussed to demonstrate how AI has penetrated modern everyday life to perform services that can be widely used to make people's lives easier [1].

6.2 FUNDAMENTALS OF ARTIFICIAL INTELLIGENCE

Programs with AI are built to function independently of following specific instructions to carry out repetitive tasks, but rather to understand the patterns of human thought and other existing ideas based on intelligent connections, thus allowing the prediction and fulfillment of the next set of instructions and ideas. They even serve to suggest business decisions and potentially richer and smarter activities for engineers, doctors, technicians, lawyers, agronomists, traders and managers, among many other professionals, always considering and updating the most relevant factors.

After a certain amount of time dealing with several analogous or not analogous cases, AI learning becomes more intuitive to make a decision. With this, AI can propose, for example, new operations based on the costs of knowledgeable people or the market, on new technologies for engineering, on previously accumulated experiences, on analyzing historical facts with coherence and knowledge, on previous budget results, on more specialized and dedicated employment plans, on market trends, on consumer demonstrations, on automatically updating catalogs, and many other activities.

DOI: 10.1201/9781003604037-6

Personal lives and knowledge accumulated by humans and their activities can serve as a database to reference AI training. There are many apps running on smartphones that are powered by AI. For this, there are many software options for interfaces, whether for a purchase on Amazon, or choosing a movie or program on Netflix, or searching for a product on Google, whose algorithms learn what the user has chosen most. Based on this information, these sites start to make suggestions, usually linked to commercial purposes, so that the user is induced to watch, buy and click a like button and so on.

In AI anyone can have many effectively functional and useful programs, such as dozens of alternative records and appointment reminders for a given day in the life of professionals or students with exams, or scheduled tasks on a given subject for an upcoming test that in daily life can come from signals controlled by AI. Very common current examples of this AI capability are the spam filters for emails in inboxes coming from social networks, which can be personalized and used to ensure that only information that really interests the user is maintained. Another very common application is defining the tastes of cell phone users in terms of the music types they usually listen to, interests in business, technological knowledge, friendships, or many other things.

AI can be loosely defined as the establishment of an ability to obtain knowledge, data and skills to apply them in various situations without user supervision. In the case of children, intelligence is usually linked to the child's personal and playful experience in their learning. There are pedagogies based on Paulo Freire and Rudolf Steiner that can still be very important. Currently, many mothers and fathers simply buy a tablet and give it as a 'gift' to their children at a young age. However, the traditional childhood learning that has always existed was based on the idea that children grow up and start to learn from their parents, siblings, teachers, friends and society in general, as well as interacting with the environment. While learning is taught, other concepts are acquired and developed naturally by children and adolescents, through their observations, correlations and personal experiences [2].

While traditional machines are known for following strict instructions, contemporary and future machines have been designed to learn, think and have the ability to perform tasks based on learned events in everyday life around them. According to John McCarthy, AI is "the science and engineering of making intelligent machines, especially intelligent computer programs" [3]. Therefore, the fundamental premise of AI is that it can create machines that think in an intelligent way that is similar to that of humans, or even better in certain technically defined aspects or in a more well thought out way. It is designed to acquire knowledge or awareness of the environment, its circumstances and entities, human or otherwise, learning whatever comes from any relationship.

AI is still a replication of how the human brain thinks, learns, decides and works when trying to solve problems, which can result in the creation of an intelligent software system [4]. The goal of AI is to achieve improvisation in the functionality of computers through functions related to human knowledge, such as problem solving, reasoning and learning. Current intelligence, as far as is known, is based on reasoning, learning, problem solving, history, perceptions, experiences and linguistic intelligence.

In order to visualize how human beings understand the functioning of their body, it is important to understand how the environmental frequencies can affect it. For this, AI is a valuable tool, as it can simulate and recreate human intelligence processes using machines or computer programs. As AI is a human creation, it is possible to establish the aspects faced by each person with logical, well-known and established explanations. In reality, AI brings together knowledge from multiple brains and it will likely surpass the thinking ability of just one individual. This chapter provides an analogy between natural and artificial intelligence as a minimum basis for understanding how the environment can influence the natural intelligence of humans and animals, by frequencies in general to which all people are subject in the modern world, regardless of their willingness.

6.3   FUNDAMENTS OF LOGICAL REASONING

The greatest interest in AI is fueled by the fact that it is a technology that can improve machines by equipping them with their own intelligence. This technology is being used for machines to help humans by teaching them how to record, adjust, adapt, query and organize more data, and process it faster, in order to always provide an answer or alternative in the best possible way. The introduction of AI into human life creates technologies that allow machines to work intelligently with or without human supervision [4].

One way to synthesize the simulation or creation of intelligent systems can be broken down into two parts. The first one is by exhibiting intelligent behavior with a built-in ability to learn, validate, explain, select, illustrate, gather data and suggest a suitable course of action for its customers. The second one is using 'helper' machines to find solutions for themselves to complex problems in a similar or better way than humans by applying compatible logic through heuristic self-learning algorithms to calculate, estimate, perform tasks for the interested party and display the necessary outputs.

To be able to make decisions, AI is made up of two main elements: an agent and its interaction with the environment. The agent can be a human or a machine or anything that can perceive its environment through sensors and act on that environment using actuators. The intelligence of agents must be calculated by their ability to create goals, how to enrich them, how to achieve them, how to select them and how to improve them. Trivial applications of AI include natural language processing, gaming, speech recognition, vision systems, memorization, traffic and more. Such applications can be commonly found in sectors such as health, teaching, science, automotive, engineering, education, commerce, communication, etc.

A basic concept for understanding AI focuses mainly on the following basic components of human intelligence: learning, reasoning, solving problems, perceiving the world and using communication languages. For this, AI and science in general use four forms of reasoning that can prove whether something can be true or not so that a definitive conclusion can be obtained:

1) By **deduction**: if A = B and B = C, then A = C;
2) By **induction**: if something is true for n = 0 and is true for n, then to be true it must be true for n + 1;

3) By **positive counter-proof** (analogy): for a rule to be true, it must be valid for all cases and if an isolated case arises, denying it without an exception statement, then this rule is not true;

4) By **reduction to absurdity** (abduction): that is, the thesis denies the hypothesis. An example of absurdity could be the observation that: "every rule has an exception; so there is a rule that says that every rule has an exception; then, the exception to this rule would be a rule without exception", thus denying the hypothesis.

Logical thinking skills play a significant role in the development of professional careers because they help a person reason through vital decisions, use histories of similar situations, generate creative ideas, set goals and solve problems. To face the challenges of daily life when entering a work sector or advancing in a particular career, an individual needs to have strong logical reasoning skills to solve daily problems in a quick, logical, reliable, competitive and creative manner. Therefore, logical thinking is the ability to think in a disciplined way or to base meaningful thoughts on evidence, learning, accumulated information and facts. The process involves incorporating logic into an individual's thinking skills to analyze a problem because it involves progressive analysis systems to find a plausible solution.

Many fields of activity, such as good project management and execution, can benefit from these logical thinking skills [5]. To be more precise, logical thinking is the act of analyzing situations and using reasoning skills and information to study a problem and reach a rational conclusion or establish alternative solutions. To be a good logical thinker, the interested party must gather all available information that can evaluate the facts proposed and methodically decide the best way to move forward with the decision to be made, establishing consequences and alternatives. Evidently, logical thinking is an essential tool for debating ideas, analyzing problems, making decisions and finding good answers at home, at work or in educational institutions.

6.4   NATURAL INTELLIGENCE VERSUS ARTIFICIAL INTELLIGENCE

AI has much to do with the natural intelligence of human beings in terms of the similar capabilities that a digital computer or a robot associated with intelligent beings have to perform tasks in a digitally controlled way. The term is often applied to projects for the development of systems endowed with intellectual processes characteristic of humans, with a similar ability to reason, discover meanings, generalize or learn from accumulated past experiences.

Analogously to the human brain, it has been experimentally and scientifically demonstrated that computers can be programmed to perform very complex tasks with great proficiency and speed. It all seems to have started with the abacus, which was an important calculating tool in very ancient societies. Around 2,400 BC, the Babylonians were already using a smooth stone instrument in addition to the records already known in India, Mesopotamia, Greece and Egypt. The benefits of the abacus can be summarized as improving attention, focus and logical reasoning to execute movement and perform mathematical operations. More recently, this abacus versus computer relationship has been ongoing since the beginning of the development of digital machines in the 1940s to the present day. Examples of this could be the demonstration of mathematical theorems, manual calculators, the definition of good business execution, psychological advice, basic calculations such as data protection in the event of invasion, the insertion of new alternatives or the running of a logical game such as chess or sudoku.

Despite continuous advances in computer processing speed and memory capacity, there are still no programs that perfectly match human flexibility in more complex and broad domains of everyday life or in tasks that require a great deal of knowledge, feelings, will or a quickly updated sensitivity. Add to this the cerebral accumulation of information that goes far beyond the present time, and may have come before childbirth inside the mother's womb, which may also have already received accumulated information and others much earlier. A greater or lesser ability of this in someone can surprise other people. On the other hand, some programs have already achieved better performance levels than human experts and professionals could expect. The execution of certain specific tasks of AI is still limited in this sense and is only found in diverse applications such as medical diagnosis, information search mechanisms on computers, the substantiation of ideas and the recognition of voice, physiognomy, fingerprints or calligraphy. It seems that everyone's imagination and sensitivity play a certain important role in this. This capacity already presents a close proximity between the information from the human brain and that coming from digital processing.

However, from certain completely different points of view, many authors dedicated to studies on natural intelligence seem to converge on the idea that it is almost impossible to have real 'intelligence' without a form of 'consciousness'. This is due to the type of human consciousness that is often conceived as an enigmatic 'mirror' of reality. Of course, this is still a phenomenon under intense investigation by science & technology, particularly in recent decades. Can thinking or conscious machines achieve what machine learning or cognitive computing seem to promise [2]? As the human brain is inside a body whose functioning is subject to environmental interference linked to chemistry, health, physics, electromagnetic waves and electricity, daily changes can occur in the information accumulated by anyone without requiring any direct control.

6.5   ARTIFICIAL INTELLIGENCE AND ITS APPLICATIONS

AI has been growing in its applications with impressive capabilities, such as those used in autonomous vehicles, planes, commercial communication or social humanoids. Although a fully functional AI-based robot does not currently exist, it only seems a matter of time before a fully functional humanoid will exist that is capable of working just like or better than people or even functioning in every home as a domestic servant or a business employee. Currently, some applications are already used regularly and are extremely important in the development of any technology that can engage in the usual activities in the same way that humans can do in real life.

Modern studies have created natural language generation (NLG), which is the part of AI that can recognize human language in the form of text and convert it in such a way as to allow a computer to communicate its data efficiently, simply and accurately. It can be used to obtain raw market data and generate reports with business summaries, or it can even generate text in readable chatbot-type interactive conversations for text-to-speech conversion. Control over the entry of spam can be carried out perfectly by NLG according to a certain classification of texts that are already widely used in similar services for analyzing the contents of emails, classifying them as true or false, with or without interest for the user.

Another important use of NLG is the classification of text to be used specifically to control spam by natural language generation. Several email providers already widely used similar NLG-based services to analyze the content of incoming emails and determine whether they are genuine or fake. To summarize, NLG can already help with the following activities:

- Suggest reliable alternatives to resolve unexpected situations;
- Analyze graphs, metrics or maps;
- Develop recurring status reports;
- Search data manually by comparing or scanning tables.

With regard to other computer activities, the following can be mentioned.

6.5.1 SIMULATION OF COMPLEX FUNCTIONS OF THE HUMAN BRAIN

Using imaging data, researchers can create complex simulations of the human brain that could lead to personalized treatments for a range of neuro-degenerative disorders and diseases. The crucial information is a complete inventory of which genes are active in which neurons, which is yet to be established. It is known that neurons are not all the same, as they are affected differently during a person's life by illnesses, temperatures, social life, personal experiences, feelings, pressure and so on, to perform alternative functions and implant different genes. The difficulties are mainly related to deducing the mixture of different neurons in various parts of the brain, recreating their electrical behavior for each cell type or even simulating how the branches of a neuron would grow from scratch [6–8].

6.5.2 COMPUTER PROGRAMMING TO USE GENERAL LANGUAGES

Computer programming still lacks the ability to truly understand human language, which includes context, textual clues, body electrochemistry effects, slang and emotional emphasis, making it very difficult to integrate all of this. Various efforts to map, understand and even reproduce this structure computationally have been underway in many countries around the world for a long time. Large groups of researchers have come together working to achieve these goals. The general understanding is how insights into brain mechanisms can help decipher the mystery of consciousness. Such an understanding could open the door to building even smarter machines than those currently available [9,10].

6.5.3 ASSESSMENT OF PROBLEM COMPLEXITY

Defining complexity as a numerical function $T(n)$ as the relationship between a time T and an input dimension n is one of the basic ideas of computer science. Algorithms based on their nature and function categorize the computational complexity of algorithms, which can be obtained as a proportion measure of the required computational resources, including time and space, consumed when a particular algorithm is being executed. An attempt is made to find heuristic algorithms that almost explain the problem and operate within a reasonable period. Algorithms such as graphs and networks, algebraic, parallel and random, are currently the most used tools [11].

6.5.4 ESTABLISHING SELF-BENEFITS

The benefits of computer tools are just an infinite incorporation of solution means for humans. Programming and coding these computers can promote and critique logical thinking, use creativity, develop persistence, develop resilience, improve children's communication skills, improve attention to detail, improve structural thinking, help in solving problems, improve mathematical skills, cause great satisfaction, help people learn to communicate better and act as desirable distraction, among many other things. Ways to learn a programming language can be found on various platforms, such as online courses, private courses, YouTube videos, books, instructors, etc. [12,13].

6.5.5 DETERMINING THE QUALITY OF IDEAS RATHER THAN EVENTS

It is possible to obtain good quality in determining the ideas to be taken into account to resolve a situation. Alternatives can be adding entertainment to the mix of ideas and events, adding value, streamlining events with well-being breaks, livening up the event with a little social media, keeping your eyes on the audience and on the screens to understand their reactions, combining content to networking, always involving everything with a new dimension. This determination of the quality of ideas should not focus too much on the event itself and thus leave good ideas aside [14].

6.5.6 SPEECH RECOGNITION

Speech recognition is a technology with which systems capture the natural conversations of a human being, analyzing the data obtained and frequency content, converting them, when necessary, into instructions to perform requested tasks. This technology has been widely used in interactive systems for voice response and mobile applications [15]. For example, nowadays almost all smartphones and desktop computers have the built-in ability to convert speech to text and vice versa.

Doctors, defense, home automation, games, criminology and even robotics also use speech recognition technology in general, each with the aim of fulfilling a specific function. However, there are times when speech recognition does not take the feelings that could be detected in words into account and therefore may not work as it should. This sometimes happens when different people pronounce the same word with different feelings or in completely different accents, even though systems are becoming increasingly adept at picking up and correcting these differences.

Another scenario where speech recognition software can fail is when there is considerable background noise, unless the microphones used are of good quality and noise canceling. The software must be able to capture the noise, analyze the frequencies of its content and generate a counter-positive noise to cancel them and, even so, it may still not be able to interpret them due to phase differences and punctual intensities.

6.5.7 FUNCTIONING AS A VIRTUAL AGENT

A virtual agent is a computer program designed to interact with humans. This type of agent can already be seen in chatbots that appear when using some company applications or when visiting some websites. For a long time now, a customer has hardly been able to detect whether they are talking to a bot or a person. With the help of AI and machine learning, the bot has been designed to respond very specifically and in a personalized way to user queries. In the contact center scenario, it is used to handle various business tasks in support of information technology (IT), where virtual agents not only perform basic functions, but can also handle complex queries based on previously accumulated situations [16].

6.5.8 ESTABLISHING DEEP LEARNING PLATFORMS

Deep learning platforms use neural networks with abstraction layers in an attempt to mimic the functioning of the human brain. They are designed to process data and create patterns from which constructive decisions can be made. These platforms are currently being widely used to identify patterns and categorize applications that are only compatible with large-scale data sets being available.

6.5.9 PREPARATION OF BIOMETRICS

Biometrics is a technology used to identify, measure and analyze the physical aspects of the structure and shape of the human body, such as the face, mood, fingerprints, temperature and specific body characteristics. This technology allows the use of more natural forms of interaction between humans and machines, including touch, fingerprints, gait, speech tone, eyes and body language recognition, as has been widely used in bank branches and in issuing documents. Biometrics are generally used to confirm a person's identity, a corresponding image or even identify commands, such as waving at a video game console system created by the US company Microsoft (Xbox) or even placing your finger on a scanner.

6.5.10 ROBOTIC PROCESS AUTOMATION

Robotic process automation (RPA) is a form of automation for business processing. In traditional workflow automation tools, a software developer produces a list of supported actions to automate a task using a dedicated scripting language. RPA systems develop a list of actions from observing the user performing that task within the software and then running automation to repeat those tasks directly within the software.

6.5.11   USE AS CONVERSATIONAL ARTIFICIAL INTELLIGENCE

Conversational AI (CAI) is a set of technologies that allow computers to simulate conversations. These technologies refer to assistant messaging applications based on human speech and chatbots to automate communication and create personalized experiences for each customer. This form of AI is preferred because it is engaging and has a time limit. In business, the conversational element keeps the customer engaged while, in the background, the user simultaneously works on the customer satisfaction aspect of the transaction or draws analogies, taking into account past experiences gained in similar operations. The number of queries that can be handled by a company can be quite large and, as such, CAI can reduce query resolution time, making it effective to resolve the issues addressed.

6.5.12   NATURAL LANGUAGE PROCESSING

Natural language processing (NLP) is an application of AI that allows machines to process and understand human language the way it was written. More recently, deep learning models can not only analyze large volumes of text, but also perform tasks such as providing a content summary, translating into other languages, modeling context and even engaging in sentiment analysis [17,18].

6.5.13   UNDERSTANDING NATURAL LANGUAGE

Natural language understanding (NLU) is a branch of NLP that converts natural language spoken by humans into structured data. It can perform two tasks simultaneously, namely intent classification and entity extraction. A user's intent is defined as what they are trying to convey or accomplish. For example, when they read a sentence, they immediately understand the meaning or intention behind it. Intent classification is a two-step process. First, an NLU model is fed with labeled data that provide the list of already known or accumulated intentions and the example sentences with analogies for these intentions. Once trained, the model acquires the competence to classify a new sentence that may fit well with one of the predefined intentions. Entity extraction is defined as the process of recognizing key information from a given text. Details like time, location, importance and user name provide context and additional information related to intent. Intent classification and entity extraction are the two main drivers of CAI [19].

6.5.14   ESTABLISHMENT OF GENERAL INFORMATION

Evidently, AI has no defined limits since everything depends on human imagination and experience. Among other general applications of AI that can be mentioned here are the creation of systems, music, data transport, space sciences, finance, hospital activities, medicine, biological intelligence, military, complex industry, video games, transportation, online consultancy and many others. Human creativity and imagination are the limits.

6.6   NEURAL NETWORKS

A simple example to understand how a computer can resemble the human or animal brain is the use of an artificial neural network. An example of a basic application of a neural network could be the use of AI to make a decision when purchasing, say, a washing machine. The neural network must be able to take into account some of the product options offered on the market that may have greater or lesser weight or importance for each specific consumer. In this example, Table 6.1 lists the aspects, x_i, taken into account with their due weights assigned by the consumer, w_i, based on the importance that this hypothetical consumer wants to give to each of them. In this case, consider the price, the quality of the brand, whether there is an acceptable maintenance network in cases of defects, whether the volume that the machine can wash at a time is compatible with the consumer's needs and whether it has a good quality guarantee of the product manufacturers.

Numerically, equation 6.1 quantifies the decision y, taking into account the aspects and weights considered in Table 6.1:

$$y = w_1 x_1 + w_2 x_2 + w_3 x_3 + \ldots + w_n x_n \tag{6.1}$$

where y is the decision total composition with the input variables x_i, each being assigned a weight w_i.

The product with the highest total to be considered in the sum y of the aspects x_i will be the final option. Table 6.2 quantifies each type of washing machine considered in the example in Table 6.1 and which ones were found on the market by the hypothetical consumer with the due weights taken into account according to the characteristics attributed to them. The sum of this table defines washing machine 2 as being the one that best meets the buyer's specifications by the highest weighted sum established by equation 6.1, which was $y = \Sigma w_i x_i = 268$.

The neural network can be represented graphically for the quantified decision to purchase the washing machine, as shown in Figure 6.1. The ramifications of this

---

TABLE 6.1
Aspects of the hypothetical purchase of a washing machine

Aspect (x_i)	Importance given	Weight (w_i)
x_1 = Price	Lowest price?	6
x_2 = Brand quality	Well recommended by users?	10
x_3 = Maintenance available	Easily available?	8
x_4 = Washable volume	Larger washable volume?	8
x_5 = Guarantee	Warranty time?	6

TABLE 6.2

Quantifying the aspects for purchasing a washing machine

Aspect = W_iX_i	Machine 1	Machine 2	Machine 3	Machine 4
x_1 = price	6 × 5.00	6 × 8.00	6 × 5.00	6 × 3.00
x_2 = brand quality	10 × 7	10 × 6	10 × 8	10 × 5
x_3 = maintenance available	8 × 3	8 × 6	8 × 4	8 × 9
x_4 = washable volume	8 × 5	8 × 8	8 × 8	8 × 3
x_5 = guarantee	6 × 10	6 × 8	6 × 7	6 × 8
Decision (y)	214	**268**	248	212

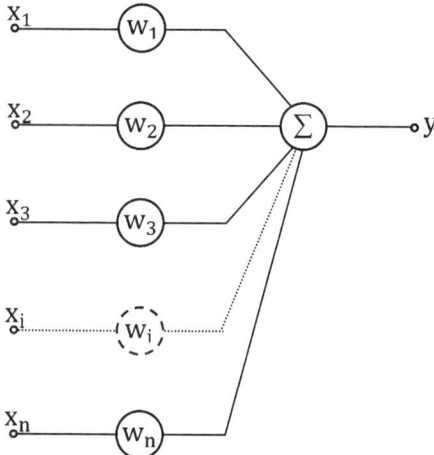

FIGURE 6.1 Basic decision-making framework in AI.

generic figure approximate a simplistic way in which the human brain could be conjecturing making a decision. Of course, the solution in this basic figure can be just one part of a final decision that is much more complex, including other generic aspects, dependences between them, special influences and the greatest relevance.

To expand the complexity of the decision a little further, just one of many possible situations related to the various aspects of a better-designed decision can be considered. It is then possible to establish other more complex situations in the final decision, such as sums y_1, y_2, y_3, ... y_i, ... y_n. That is, the neural network can take into account many aspects of the same problem and then reach a much more solid conclusion.

To establish the complexity of information that can be taken into account for a decision, according to *Scientific American* [20], the memory capacity of the human brain has the equivalent of 2.5 petabytes (2500 TB) and is composed of 100 billion nerve cells called neurons. This number was obtained by estimating the amount of information that can be stored by the 125 trillion synapses in the cerebral cortex, which are connected to another thousand cells by axons. Stimuli from the external environment or input from sensory organs are accepted by dendrites that create electrical impulses that can travel quickly through the neural network.

According to Chapter 4, to understand better how brain communication works, it is necessary to take into account that a neuron has the ability to send a message electrochemically to another neuron and thus form a network with the capacity to deal or not with a given problem. Artificial neural networks (ANNs) are made up of several nodes, which imitate the biological neurons of the human brain and are capable of learning what is happening around them, changing the values of each weight and adapting themselves to each new situation. Neurons are connected by links and interact with each other. Nodes can receive input data and perform both simple operations with these data and, beyond that, more complex interactions. The result of these operations is passed to other neurons and so on. The output at each node is called the node's activation or value. Each link is associated with a weight. With the AI data obtained from neural networks, simple operations or ones that are much more complex can be carried out [21–23].

REFERENCES

[1] Fundamentals of Artificial Intelligence, Module 1: The Imagine Cup Junior guides and lesson materials created by Microsoft and their partners and intended to be for guidance only to support with the Imagine Cup Junior Challenge. For the latest on Microsoft AI, please, visit www.microsoft.com/en-us/ai.

[2] R. Penrose, E. Severino, F. Scardigli, I. Testoni, G. Vitiello, G. M. D'Ariano, and F. Faggin, Artificial intelligence versus natural intelligence, in Fabio Scardigli (ed.), Philosophy and Religion (R0), 1st ed. Springer, https://doi.org/10.1007/978-3-030-85480-5, ISBN978-3-030-85479-9, 25 Mar 2022, p. 192.

[3] J. McCarthy, What Is Artificial Intelligence? Computer Science Department Stanford University, http://www-formal-stanford.edu/jmc/, pp 1-15, 12 Nov 2007.

[4] L. Bermudez, Overview of artificial intelligence buzz – machine vision – medium [Online]. Medium, 2017. Available at: https://medium.com/machinevision/overv iew-of-artificial-intelligence-buzz-adb7a5487ac8. Accessed 29 Jun 2019.

[5] M. Narang, What is logical thinking – Significance, components, and examples, Knowledge Hut, www.knowledgehut.com/blog/career/what-is-logical-thinking, 13 Jul 2023.

[6] R. Ananthanarayanan, S. K. Esser, H. D. Simon, and D. S. Modha, The cat is out of the bag: Cortical simulations with 10^9 neurons, 10^{13} synapses. In Proceedings of the Conference on High Performance Computing Networking, Storage and Analysis (SC '09), Article 63, 1–12. https://doi.org/10.1145/1654059.1654124, S2CID 6110450, 14 Nov 2009.

[7] M. Colombo, Why build a virtual brain? Large-scale neural simulations as jump start for cognitive computing, Journal of Experimental & Theoretical Artificial Intelligence, 29(2), 361–370. https://doi.org/10.1080/0952813X.2016.1148076. S2CID 205634599, 04 Mar 2017.

[8] C. Eliasmith, T. C. Stewart, X. Choo, T. Bekolay, T. DeWolf, Y. Tang, and D. Rasmussen, A large-scale model of the functioning brain, Science, 338(6111), 1202–1205. https://doi.org/10.1126/science.1225266, PMID 23197532, S2CID 1673514, 30 Nov 2012.

[9] M. Warschauer and D. Healey, Computers and language learning: An overview, Language Teaching, 31, 57–71, 1998.

[10] L.T. Collins, Towards simulating the human brain, Influence: Science in Service to Society, 19(3), Art 2, 2017.

[11] A. Kuzmiakova, Computer Science, Algorithms and Complexity, Arcler Press, p. 254, ISBN-10: 1774077485, ISBN-13: 978-1774077481, 01 Nov 2020.

[12] M. Colombo, Why build a virtual brain? Large-scale neural simulations as jump start for cognitive computing, Journal of Experimental & Theoretical Artificial Intelligence, 29(2), 361–370, 04 Mar 2017.

[13] R. Mailler, J. Avery, J. Graves, and N. Willy, A biologically accurate 3D model of the locomotion of *Caenorhabditis elegans*. In 2010 International Conference on Biosciences. https://doi.org/10.1109/BioSciencesWorld.2010.18, ISBN 978-1-4244-5929-2, S2CID 10341946, 7–13 Mar 2010, pp. 84–90.

[14] B. Matthew Kudrowitz and D. Wallace, Assessing the quality of ideas from prolific, early-stage product ideation, Journal of Engineering Design, 24(2), 120–139. https://doi.org/10.1080/09544828.2012.676633, 2013.

[15] Motor Harley Davidson Co., Livewire Initial Care, Service Bulletin M1502: 2020, 03 Oct 2019.

[16] The Amelia Blog, Blog Digital Workforce Summit 2019, IPsoft, Microsoft 2023, 25 Feb 2019.

[17] L. Tunstall, L. von Werra, and T. Wolf, Natural Language Processing with Transformers: Building Language Applications with Hugging Face, 1st ed. O'Reilly Media, ASIN: 1098103246, ISBN-10: 9355420323, ISBN-13: 978-9355420329, 01 Mar 2022, p. 406.

[18] T. Wolf, L. Debut, V. Sanh, J. Chaumond, C. Delangue, A. Moi , P. Cistac, T. Rault, R. Louf, M. Funtowicz, J. Davison, S. Shleifer, P. von Platen, C. Ma, Y. Jernite, J. Plu, C. Xu, T. Le Scao, S. Gugger, et al., Transformers: State-of-the-art natural language processing. In Proceedings of the 2020 Conference on Empirical Methods in Natural Language Processing: System Demonstrations, 2020, pp. 38–45.

[19] M. Mantha, Conversational AI: Design & Build a Contextual AI Assistant [Online]. In Fundamentals of Artificial Intelligence – Module 1, Imagine Cup Jr, Student Guide. Available at: https://towardsdatascience.com/conversational-ai-design-build-a-contextual-aiassistant-61c73780d10, 2019.

[20] P. Reber and D. Smith, What is the memory capacity of the human brain? Scientific American Mind, 01 May 2010.

[21] S. Herculano-Houzel, The human brain in numbers: A linearly scaled-up primate brain, Frontiers in Human Neuroscience, 3, 31. https://doi.org/10.3389/neuro.09.031.2009, PMC2776484, PMID: 1991573119915731, prepublished online 05 Aug 2009. Published online, 9 Nov 2009.

[22] M. Negnevitsky, Artificial Intelligence, A Guide to Intelligent Systems, 2nd ed. Addison-Wesley, 27 Oct 2021, p. 435.

[23] J.L.G. Rosa, Fundamentos da Inteligência Artificial, Instituto de Ciências Matemáticas e de Computação-USP-Br. LTC, www.grupogen.com.br, http://genio.grupogen.com.br, ISBN 978-85-216-0593-5, 2011.

7 Id, Ego and Superego

7.1 INTRODUCTION

The dimensions (volume) of human brains are practically the same for adults and it seems impossible that anyone could have a much greater capacity to store information than others. Managing this information sets people apart. It really seems that the information within the human brain can be allocated to the ego, the superego and the id. The superego (morality) is what the person is using at a given moment to think or communicate with others to perform some task. The ego (reality) is what is available to access consciously what is needed at the right time. The id (instincts) are the 90% that cannot be easily accessed because the information that people receive throughout their lives is stored there, without them being fully aware or in control of it [1].

7.2 MENTAL INFORMATION

People receive much of the information contained in the id without realizing it and some of it is received even inside the mother's womb, and it is stored unconsciously throughout life. The id creates the demands, the ego adds the needs of reality and the superego incorporates morals into the action. The results of the more radical elements of these influences make an individual at peace with themselves even if, at times, their mind has to resort to shields and disguises to get there.

The superego is the moral aspect of the individual's personality according to Sigmund Freud's theory of psychoanalysis and is responsible for 'taming' the id, which is repressing primitive instincts based on moral and cultural values acquired throughout life. The superego is the rational part of the brain, but it can be highly influenced by instincts if it is weak or has little moral or rational training. The id is the instinctive part (without reasoning) where the ideas received throughout life are irrationally stored, from conception inside the mother's womb to the present. The id simply stores data, without logic or criteria, and the superego uses it as it sees fit. The id is instinct, the superego is reason and the ego is communication with the world, as represented in Figure 7.1. To summarize, the id, ego and superego are the three components of personality that represent, respectively, impulsiveness, rationality and morality.

 DOI: 10.1201/9781003604037-7

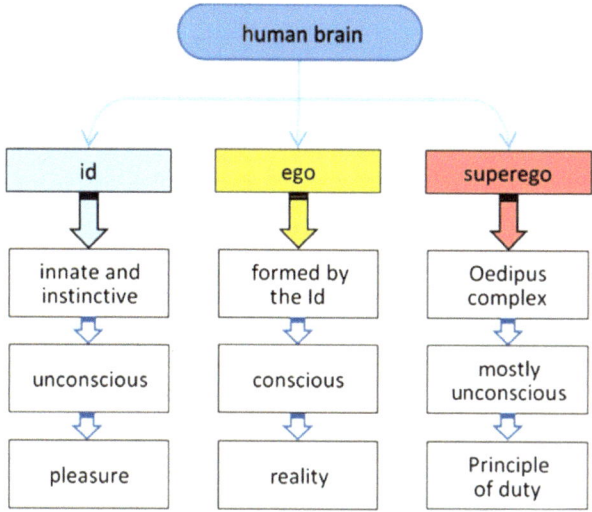

FIGURE 7.1 Constitution of the human brain.

The ego is as if it were the manager of the information contained in the id. The superego is the organizer of logic, common sense and the establishment of prejudices, in addition to being the controller of information, controlling both the information that enters and leaves the ego and also how it enters and leaves the id. The superego has learned throughout life everything it needs to do to keep the body alive, defensively or as an attacker. It is the neuronal experiences recorded and contained in neurons that establish defenses related to pain, peace, joy, extreme noises, sadness, danger, moments of attack, moments of defense … in short, everything else that comes together in human feelings in a lifetime.

It is interesting to say that what is called subconscious reasoning is the processing (combination) of information already contained in the id without having a conscious notion of it. People then think that it was an 'angel' who informed them, for example, that they had an illness or that someone inconvenient was waiting for them somewhere around, or that they would be robbed, or that something bad would happen to them.

The Oedipus complex is a phenomenon described by Freud to designate when the child will develop loving feelings towards the parent of the opposite sex and hostile feelings towards the parent of the same sex. In reality, all of this was just processed unconsciously by the brain based on information, deductions that were already there that were incorporated here and there throughout life, and that unconsciously were putting together what could happen to the person next. Believers call the soul the origin of this information stored in the terabytes of the brain. Therefore, this soul would disappear after the body is dead. Spirit would be what is recorded or witnessed in the bearer's brain and in other people's brains or in things they have had contact with throughout their lives by exchanging ideas with them. Therefore, the spirit would never die.

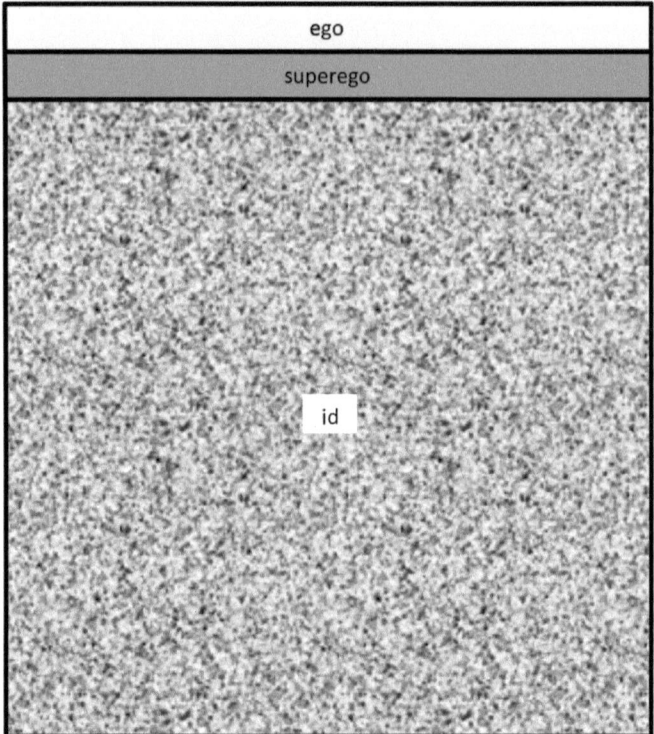

FIGURE 7.2   Freud's proportional view of the human mind [4].

Figure 7.2 shows a mental representation of human mind occupation as an iceberg according to Sigmund Freud [2]. This is only an illustration of the current geometric equivalent of what is known about brain occupation from information captured throughout life to the present.

Humans have the largest brain in proportion to body size of any other living creature. This appears to be related to a greater capacity for storing data and knowledge in memory. As explained in Chapter 1, this accumulation of human experiences in the world is based on information received through a combination of the fundamental senses: sight, touch, hearing, taste and smell. The memory of a lunch someone had at a restaurant with friends a few days ago, for example, may include the taste and smell of the lasagna, the richness of the salad mix, the attentive service of the waiter, the background music, the presence of some people known, or the topics of conversation. These components of the human experience can activate various parts of each person's neocortex. However, the episode itself would initially be stored in the hippocampus and this memory is consolidated over time. The long-term storage of this experience will be distributed across different parts of the neocortex. Once a memory is stored, the hippocampus becomes critical, serving as a memory index for locating the information contained therein [3].

Returning to the story of the restaurant lunch, when one of the friends recalls the occasion and how much the friends enjoyed that environment, the interior of the restaurant comes to mind, causing each person's visual cortex to relive the situation. The part of the brain that receives and processes sensory nerve impulses from the eyes becomes active with a pattern similar to that when they arrived at the restaurant. Due to synaptic plasticity and strengthened connections, this visual seed becomes sufficient to access the lunch scene with friends in the index of hippocampal memories. The hippocampus then directs the situation in neuronal traffic back to the appropriate circuits of the neocortex, reactivating the environmental sound, the delicious taste of the food, the topics of conversation, the service of the helpful waiter and any other components that were part of that occasion. All of this is still conjecture about how memory indexing and recall happens [4].

An analogy of the good functioning of storing experiences when everything works well is to consider memory as a digital database or even an old-fashioned office file folder. Any current event will trigger a search in the database and the memory of what happened will be recovered.

The two strongest types of learning are visual and auditory, contradicting the idea that different people have different learning styles. In fact, this is a widespread misconception. A 2012 survey of teachers in the Netherlands and the United Kingdom, for example, found that more than 90% believe in different learning styles. Even if people could have different personal preferences regarding the way they capture information, it is not true that an individual can learn best through any specific set of sensory cues.

In 2008, the cognitive psychologist Dr Harold Pashler and colleagues at the University of California, San Diego, who were evaluating decades of research, conducted an extensive review of learning. They found no evidence to support the idea that an individual can learn more effectively when teaching is adapted to a specific style. The mistake they make "is based on a valid research finding that visual, auditory and kinesthetic information is processed in different parts of the brain", as the authors of the UK–Netherlands study explained. It is known that during learning, as in many other brain activities, these brain areas interact with each other and not in isolation. However, generally speaking, there is no evidence that these areas work 'better' or 'worse' in some people than in others to determine how each of them can learn better [5].

A person is more likely to remember something they learned if they have some emotional attachment to a certain fact associated with the positive side of their stress and confusion. This happens because the amygdala stimulates memory, improving attention and perception, and can help retain this experience in memory, triggering the release of stress hormones. Researchers at The Neuroscience Research Institute of the Queensland Brain Institute (QBI), University of Queensland, Australia, discovered that bad experiences automatically improve the formation of a memory about the places experienced and can serve as a warning to avoid similar potential threats [1]. On the other hand, too much stress can overload memory, cause anxiety and impair knowledge. With this, it was concluded that there is a right amount of information that can optimize alertness and cognitive performance. These scientists' research has also shown that being confused by new ideas, or about an unexpected situation, can encourage a person to work harder to understand, thus leading to a deeper understanding and better retention of what they have learned [6].

An interesting fact about visual information is that there is much evidence showing that the blue light emitted by smartphones, tablets and computers suppresses melatonin production [7]. This fact indicates that using these devices at night can interfere with the natural cycles of sleep and the body in general. Sleep is known to be important for learning and is crucial for consolidating long-term memories. On the other hand, lack of sleep can also impair attention and short-term memory. The teenage brain is particularly sensitive to the effects of blue light, leading experts to recommend that teens avoid nighttime use of devices that emit blue light if they want to sleep better.

As anyone can see, many conditions already known in medicine can affect the health and performance of the brain and, therefore, its ability to learn. This includes the conditions experienced by mothers, both present and before birth, to include genetic disorders, diseases and injuries acquired throughout life. The effects can be temporary, such as a concussion, for example, which occurs when the brain hits the inside of the skull in a sudden head movement, causing short-term symptoms, including memory loss and difficulty concentrating. Hearing or speech impairments, especially if unrecognized, can affect communication in the classroom. Other conditions, including Alzheimer's disease, can affect brain areas crucial for learning and cause irreversible damage as far as is known.

7.3   BRAIN EVOLUTION FROM CHILDHOOD

The human brain is loaded from birth, and even before, with an enormous amount of information coming from the most varied sources and recorded in the feelings and experiences lived by the mother. Among these are sound, visual, sensory, vibrational and gustatory sources, which add to each person's imagination and experience in greater or lesser quantities. It happens that people with headaches, fever, stomach upset or any other organic disorders change their spectrum of colors and heat around the body, better known as their aura, which can indicate what is happening to them. As is to be expected, some people since childhood have the physiological capacity to observe more or less broad intensities of colors, sounds or heat around things and other people that will reflect the physical characteristics or ailments in the colors they perceive. Therefore, as children grow up with this sensitivity, they associate the heat and colors of other people's auras with human ailments differently from ordinary mortals. Other people are surprised by this ability, which is usually recognized in some religions as mediumship or clairvoyance, but which seems to be nothing more than the perception of a wider range of colors compared to ordinary mortals.

As discussed in Chapter 5, similar to people capable of perceiving the aura, maintenance teams at electricity companies use detectors or cameras sensitive to infrared radiation (IR) to locate heat caused by possible defects in electrical wiring connections. In addition to these, several other devices also work using IR radiation, such as remote controls. These controls use a light-emitting diode (LED) that emits IR radiation so that it is then detected by a sensor in the electronic device, as is the case with automatic doors, televisions, barcode readers and computer mice. It is clear that the human body also manifests a certain more or less intense heat around itself and, therefore, infrared radiation.

A connectome is a map that brings together neural connections in the brain of living beings, and can be thought of as its 'connection scheme'. As is known, any nervous system of a living organism is made up of neurons that communicate through synapses. Therefore, image connectomics is based on graph theory and is an effective and unique methodological framework for studying brain functional connectivity patterns in development and aging. It is known that normal brain development is characterized by the continuous and significant evolution of the neural network during early childhood and then adolescence, following specific patterns of maturation. Normal aging is related to some changes in resting state brain networks, which are associated with a certain cognitive decline over time.

Tracking the evolution patterns of the connectome throughout life is a great challenge to be able to design an integral metric such as its complexity to understand the principles of network organization in the human brain. In the study of this topic, a brain network autoentropy (NEE) was initially defined based on the energy probability (EP) of each brain node. In fact, NEE comes from the French *née*, the feminine past participle of *naitre*, 'to be born'. In one study [1], NEE was used to characterize the life order trajectory of whole-brain functional connectivity in 173 healthy individuals aged between 7 and 85 years. The results revealed that throughout life, whole-brain NEE exhibited a significant nonlinear decrease and that the distribution of EP changed from a certain concentration to a wide dispersion, implying an increase in the order of the functional connectome over age. Furthermore, the brain regions with significant changes in EP from the flourishing period from ages 7 to 20 to the juvenile period from 23 to 38 years were mainly located in the right prefrontal cortex and basal ganglia. These changes were involved in emotional regulation and executive function in coordinating previous knowledge with the action of the sensory system. This implied that self-awareness and voluntary control performance changed significantly throughout neurodevelopment. However, the changes from youth to middle age, from 40 to 59 years old, were located in the mesial and caudate temporal lobe, which are associated with long-term memory, indicating that the memory of the human brain begins to decline with age during this period. Overall, findings related to this phenomenon suggested that the human connectome changed from a relatively anatomical state to an ordered and organized state with a lower entropy [8].

Since its inception, social neuroscience has sought to clarify the formation of impressions of people and groups by using representations to guide their actions. Recently, these complex forms of social cognition and efforts to do so have been reinvigorated with computational modeling. It can provide a framework that delineates specific processes underlying social cognition and relates them to neural activity and behavior. In the literature, one can find primers on computational modeling describing how it has been used to elucidate the psychological and neural mechanisms of impressions, social learning, decision-making, morality and intergroup prejudices [9].

7.4 MANAGEMENT OF BRAIN INFORMATION

An interesting aspect of the accumulation of information is the brain's ability to memorize what happens in the intestines, ears, eyes and other human organs and senses. From birth or even before, much information is received and accumulated from the environment in which people live through the senses. The storage of all this data in the human brain must be more or less within 100 Tbytes (1 Tbyte = 1,024 gigabytes $=10^{12}$ bytes) of neurons. Only something like 10% of this can be used by the conscious human mind (ego and superego). Of note, some IT companies have already approached this and announced the production of solid-state drives (SSDs) with a capacity of 100 Tbytes or more. When locating information, having all this accumulated in human memory, sometimes the brain, or its 'search manager', presents certain difficulties in finding an immediate solution to a certain unusual situation. Therefore, the search for data within the brain can be rapid because it starts randomly and occasionally with the desired information, or with the most recent information, or with the small amount of information available. Otherwise, it may be time-consuming. It thus becomes evident that the enormous amount of information accumulated throughout life makes it difficult for the memory manager to find it immediately among the abundant information accumulated since a young age. This amount of accumulated information perhaps explains why elderly people cannot immediately remember some things they are asked or what they want to say. After some time, these memories emerge unexpectedly, often when the subject is no longer of interest in the conversation.

Another interesting aspect of memory is the information processing contained in the human subconscious that makes a person think that it was 'someone' outside the physical world who informed him or her about one or another unexpected situation that was not consciously foreseen. For example, someone who does not want to get somewhere is subconsciously alerted by the abundant unconscious information contained in the brain. Information accumulated subconsciously over time tells this person that there will be a complicated problem to be faced there. If this someone 'outside' the physical world is contradicted, there will most likely be unpleasant facts to be faced. A situation contrary to this can also happen, that is, the subconscious informs a person that the best solution to a certain problem will be to go to a certain place as defined in the subconscious and the problem will be easily solved.

Cultivating positive ideas always helps the subconscious to carry positive solutions to daily problems and vice versa. These positive ideas must be very clear; for example, avoid thinking, "tomorrow I will be healthy", as this can continue for a long time afterwards, always waiting for tomorrow. Convincing the subconscious that "I am healthy" can now predispose the best use of the information already contained in the brain. Be careful not to ask for things like, "I want to make a lot of money", as this can be very expensive to do! It may happen that there is an accident causing the loss of a leg or an arm and the insurance company having to make a prompt payment of millions.

Prayers such as those preached by religions seem to be ancient wisdom to convince the brain that one must have good ideas to have good solutions and help. People who preached repetitive prayers since ancient times probably had not initially understood the reason for this cerebral need to preach positive ideas, but simply repeated it without criticism and, perhaps in a poeticized way, what they had learned over time until today.

The brain makes up about 2% of a person's weight, but uses 20% of its oxygen and calories for hydration, as was first established in 1945, when scientists estimated that the brain is made up of about 73% water. Keeping a brain hydrated is important. Dehydration by just 2% can impair a person's ability to perform tasks involving attention, memory and motor skills [10].

Cholesterol is a type of fat that people often consider bad for their health. It is true that eating too much cholesterol is bad for your heart. Approximately 25% of the body's reliable source of cholesterol is contained in brain cells. Due to the complexity of these organs, scientists are still learning about it. However, many people are unaware that cholesterol plays a significant role in their brains, as without cholesterol, brain cells would not survive.

The notion that a person only uses 10% of their brain is a myth. Functional magnetic resonance imaging (fMRI) is a neuroimaging technique used to study human brain reactions while a person performs some activity. fMRI scans show that even simple activities require almost the entire brain to be active. Although there is still much to learn about the brain, researchers continue to fill in the gaps between fact and fiction.

7.5   HUMAN COMMUNICATION

It must be said at this point that everything contained in the brain is only processed unconsciously based on information and deductions received here and there throughout life in addition to clues that were unconsciously assembling what could happen next. Believers call this information stored in the terabytes of the brain the 'soul'. The 'spirit' seems to be what is recorded in the external environment, in one's own brain and in that of other people with whom they interact. There is an exchange of ideas witnessed with them in such a way that when the individual logically dies, these ideas will remain in these people's heads. As a result, this knowledge is 'immortal'.

It is interesting to mention here that the writer Louise Hay describes in her first two books her beliefs, associating them with an emotional pattern. If this pattern is specifically negative, it may be directly linked to the emergence of physical illnesses such as cancer, where curing the causative pattern could consequently cure the illness. In her 1984 bestseller, this author also highlights her view on AIDS: "I believe that sexually transmitted diseases are almost always linked to sexual guilt. This stems from a feeling, often unconscious, saying that it is not right to express our sexuality. A person with a sexual disease can have several partners, but only those with a weaker emotional and physical immunity will be susceptible to the disease" [11].

Possibly, when Jesus started talking about the brain to men at that time in the Bible, they were revolted and offended because they did not know what it was really about. Jesus then said that whoever governed everything was on high (the Most High), which could actually mean at the top of the head. However, people at that time understood the soul or spirit as coming from heaven. Phrases of Jesus that reinforce this interpretation could be:

Have faith and you will remove mountains.

I am who I am.

Wherever you are there I will be with you, no matter the distance, we will never be far from each other.

In addition, you will know the truth and the truth will set you free.

For God so loved the world that he gave his only begotten Son, that whoever believes in him should not perish but have eternal life.

Father, forgive them, because they do not know what they are doing.

I am the resurrection and the life. Whoever believes in me, even if he dies, will live; and whoever lives and believes in me will never die.

What good is it for a man to conquer the whole world if he loses his soul?

Love your enemies, do good to those who hate you, bless those who curse you, pray for those who mistreat you. If someone hits you in the face, turn the other cheek.

Bad advice can ruin a day, a year or an entire life.

Whoever wants to be a leader must first be a servant. If you want to lead, you must serve.

The mouth speaks of what the heart is full of.

The Lord hides some things from the wise, but reveals them to little ones.

Fear not, I have conquered the world!

Nevertheless, seek first the Kingdom of God and His Justice, and other things will be added to you.

Do not judge according to appearance, but by righteous justice.

For what does it profit a man if he gains the whole world and loses his soul? Alternatively, what will a man give in exchange for his life?

Love one another, as I have loved you.

Love your neighbor as yourself.

The eye is the lamp of the body. If your eye is good, your whole body will be filled with light. However, if he is evil, your whole body will be filled with darkness. If the light in you is turned off, the darkness is immense.

A thousand will fall at your side, ten thousand at your right. However, you will not be hit.

Do not do to others what you do not want others to do to you.

REFERENCES

[1] F.A. Petocz, Psychoanalysis and Symbolism. Cambridge University Press, ISBN: 9780521021500, 2005, p. 300.

[2] C.D. Green, Where did Freud's iceberg metaphor of mind come from? History of the Psychology, American Psychological Association, 22(4), 369–372. https://doi.org/10.1037/hop0000135_b. PMID: 31633371. Nov 2019.

[3] C. Mutti, F. Misirocchi, A. Zilioli, F. Rausa, S. Pizzarotti, M. Spallazzi, and L. Parrino, Sleep and brain evolution across the human lifespan: A mutual embrace,

Frontiers in Network Physiology, Sec. Networks in Sleep and Circadian Systems, 2, https://doi.org/10.3389/fnetp.2022.938012, 03 Aug 2022.

[4]   J.L. Oschman and M.D. Pressman, An anatomical, biochemical, biophysical, and quantum basis for the unconscious mind, International Journal of Transpersonal Studies, 33(1), https://doi.org/10.9769/EPJ.2013.5.1.JLO.MDP, 12 Dec 2015.

[5]   R. Sternberg and K. Sternberg, Cognitive Psychology, 7th ed. Thomson Wadsworth, a part of The Thomson Corporation. Thomson, the Star logo and Wadsworth, ISBN10: 1305644654, ISBN13: 978-1305644656, 2006.

[6]   C. Barry, F. McMillan, and A. Woodruf, The Brain Series, Issue Two, How Memories Are Made, Chapters 1, 2 and 3. Queensland Brain Institute, Science of Learning Research Centre, The University of Queensland, Apr 2020.

[7]   A. Shechter, E.W. Kim, M.P. St-Onge, and A.J. Westwood, Blocking nocturnal blue light for insomnia: A randomized controlled trial, Journal of Psychiatry Research, 96, 196–202. https://doi.org/10.1016/j.jpsychires.2017.10.015. Epub 21 Oct 2017, PMID: 29101797, PMCID: PMC5703049, 2018.

[8]   Y. Fan, L. Zeng, H. Shen, J. Qin, F. Li, and D. Hu, Lifespan development of the human brain revealed by large-scale network eigen-entropy. College of Mechatronics and Automation, National University of Defense Technology, Entropy, 19(9), 471. https://doi.org/10.3390/e19090471, 04 Sep 2017.

[9]   L.M. Hackel, D.M. Amodio, Computational neuroscience approaches to social cognition, Current Opinion in Psychology, 24, 92–97, https://doi.org/10.1016/j.copsyc.2018.09.001, Dec 2018.

[10] M. Torrijos-Muelas, S. González-Víllora, A.R. Bodoque-Osma, The persistence of neuromyths in the educational settings: A systematic review, Frontiers in Psychology, Sec. Educational Psychology, 11, 12 Jan 2020.

[11] L.L. Hay, You Can Heal Your Life [Internet Archive]. Hay House, 1987.

8 The Power of the Mind

8.1 INTRODUCTION

Voluntary manifestations of the mind are sensitivities that people have to a greater or lesser extent to subconsciously observe other people's physical signs or mental vibrations through visual contact, knowledge, touch, physical colors or other natural means. This could be considered as a quality or defect that some people have for extra perception of the conventional frequencies under which the human being is immersed throughout his/her life and contacts. With these abilities, the active or passive subconscious, one's own or others, can be accessed through hypnosis, telepathy, radioesthesia, intuition, spells, mediumship, visions and precognition, as well as other less widespread forms, such as telekinesis, clairvoyance, precognition and teleportation. This is the subject of this chapter.

8.2 VOLUNTARY OR NON-VOLUNTARY MANIFESTATIONS OF THE SUBCONSCIOUS

Important transitions in intelligence levels do not need to be sudden, inevitable or dramatic, and nor do they necessarily change immediately in relation to what an organism has the capacity to do, as they change this capacity for evolution in descendant lineages. Of note, the most useful way to consider changes in cognitive evolution is to consider the selection action in current computational architectures that also support cognition. The richness of the subsequent evolutionary divergence of living beings allows the exploration of the possibility of a great transition of differential wealth that has not yet been thought of. Exceptional selective forces, such as selection for efficiency, stability or robustness, can drive the evolution of structural changes in the current computational architecture, which in turn alter the evolutionary capacity of the same lineage. Thinking of evolution in terms of major transitions therefore provides a broader perspective that emphasizes the gradual changes in possible evolution.

DOI: 10.1201/9781003604037-8

8.2.1 HYPNOSIS

Hypnosis is a special psychological state induced through various alternatives, such as using a pendulum, finger movement or simply vocal means. It happens due to the sensitivity that the hypnotist has to access susceptible points in the other's brain and influence them. This technique has recently been increasingly used as an instrument in different diagnoses. It should be noted that hypnosis is an altered state of consciousness that allows others access to the subconscious where emotions, feelings, desires, habits and long-term memories are found that people cannot control at the conscious level. According to the American Psychological Association (APA), hypnosis is a procedure in which a health professional or researcher treating someone suggests that they will experience changes in their body's sensations, perceptions, thoughts or behavior. This consists of taking the person into a trance state, increasing their focus and enabling them to know themselves better, remember past situations in detail, give a new meaning to traumas, or overcome fears or physical or psychological ailments [1].

Hypnosis can be effective in many cases. It must be said that it is a very natural process, having nothing mysterious or new about it, as it is just a state of trance. It is nothing more than a state common to the human body, since the person himself/herself can enter a trance of this type alone if this person knows how to do it. Hypnosis is just relaxation and imagination used by professionals who hypnotize their patients. In this state of deep concentration, hypnotized people are able to remember situations from the past and work their mind to induce healthier habits, among other purposes. Such a trance state is normal for the human body of people who often also practice self-hypnosis and are able to enter this state on their own.

Hypnosis had a 93% success rate in patients after only six sessions in treatments that could last a month or two or so. This shows that hypnosis, in addition to being very effective and successful in treating the majority of patients, can also help them recover in a much shorter time compared to other researched types of therapy. In hypnotic trance, the patient can be led to immerse themselves within themselves, making them remember, for example, the origin of the traumas they are currently experiencing, probably caused by childhood situations. It is also possible to understand why a person acts the way they do in different situations and uncover the triggers and reasons for their behavior and habits. In addition to being hypnotized, the patient can learn self-hypnosis, allowing them to enter a trance state again and get to know their subconscious and themselves better.

Another great benefit of hypnosis is that it can be used effectively to treat different types of problems, both physical and psychological, such as depression, anxiety, phobias and addictions, physical pains, weight loss, sexual dysfunctions, personal development and improvement in memory, among others. It is important to emphasize that the purposes of hypnosis are always therapeutic and that just because it can be used for different purposes does not mean it should be applied for entertainment. Experts make it clear that TV shows with people being 'hypnotized' to be exposed to ridicule are not part of professional hypnosis.

If hypnosis is applied by a professional or by self-hypnosis, the hypnosis process does not have any side effects, which is very different from other types of psychological and psychiatric treatments, which use medications that can have side effects. Hypnosis is a completely natural process, as the only tool used during the process is the human mind itself. No patient needs to take medication to be induced into a trance state. As it is a natural process with no side effects, hypnosis is often used as a complementary treatment in patients with psychological or psychiatric problems.

8.2.2 TELEPATHY

Telepathy is the ability to perceive emanations from another's brain associated with information transmitted by manifestations of patients and the telepath's sensitivities. Knowledge or communication may be direct or at distance between two minds by someone whose mental processes are sensitive to signals beyond the limits of ordinary human perception.

Telepathy is the direct transfer of thoughts from one person (sender or agent) to another (receiver or perceiver) without using the usual sensory channels of communication and is therefore a form of extrasensory perception (ESP). Although the existence of telepathy has not yet been proven, some parapsychological research studies have produced favorable results using techniques such as card divination with a special deck of five sets of five cards. The agent can simply think of a random order of the five card symbols, while the percipient tries to think of the order in which the agent is focusing. In a general ESP test, the sender focuses on one card face at a time, while the recipient tries to think of the symbol assigned to it. A screen or some greater obstacle or distance, of course, separates both subjects. Scores significantly above chance are extremely rare, mainly because testing methods have become more rigorous [2].

Findings from the telepathy study suggest an association between telepathy and the right parahippocampal gyrus [3–5]. The methodological rigor, the isolated and robust brain activation with telepathy, and the established theoretical relevance of this brain region with reference to paranormal phenomena highlight the need for further studies using advanced fusion imaging techniques (simultaneous functional magnetic resonance imaging (fMRI), electroencephalography (EEG) and magnetoencephalography (MEG)) so that telepathy can be determined as a science. There is some disagreement as to how new this approach really is. *IEEE Spectrum* reports that this recent study is quite similar to one already carried out at the University of Washington. In that study, researchers used the same EEG setup for transcranial magnetic stimulation (TMS), but instead of pulsed light, they stimulated the brain's motor cortex to cause the recipient to unconsciously press a key on a keyboard. Pascual-Leone, a Spanish-American professor of neurology at Harvard Medical School, USA, however, states that his work is notable because the recipient was aware of the communication [2]. Both studies represent just a small step towards telepathy engineering that may still take years or decades to perfect. In a more rigorous analysis, the goal is to remove the intermediary computer from the transmission equation and allow direct brain-to-brain communication between people [6,7].

Outside of medicine, brain-to-brain communication can find applications in many disciplines. Soldiers, for example, could use the technology on the battlefield, sending commands and warnings to each other. Civilians can also benefit; businesspeople could use it to send tips to partners during negotiations, or pitchers and catchers could prevent sign stealing during baseball games. Still, telepathic communication works like a kind of futuristic walkie-talkie involving major advances in detection, emission and reception technologies and perhaps even a slight retraining of the human brain. At the same time, Pascual-Leone warns that scientists must also keep in mind telepathy ethics [3,7–9].

8.2.3   DOWSING OR RADIOESTHESIA

Dowsing, rabdomancy or radioesthesia is the practice of divination with the aid of a wand or a fork known as a magic wand or a witch's wand. The word has Greek origins, being the combination of *rhabdos*, which means stick or rod and *manteía*, which means divination. The person who practices dowsing is called a dowser. This designation seems to have biblical origins, as in the Holy Bible, Hosea 4:12, it says "My people inquire of a piece of wood, and their walking staff gives them oracles. For a spirit of whoredom has led them astray, and they have left their God to play the whore". Radiesthesia is also associated with searching for hidden objects and sources of water with the help of a magic wand or stick.

Radiesthesia was practiced in New Spain by the popular classes. Examples of this activity were recorded in the 17th and 18th centuries in accusations presented to the Inquisition. Although it had some adherents, it was often considered ridiculous. What is interesting, however, is the fact that this widespread opinion coexists with a religious sight that viewed radiesthesia as a sin. The evidence suggests, however, that the first view prevailed over the second, so that dowsers were often mocked but never punished. Many articles seek to explain the phenomenon of radiesthesia within this context and interpret the ways in which this "secret and science" was practiced [10].

Uri Geller, born Gellér György on December 20, 1946, is an Israeli, naturalized British, who became famous in the 1970s, at the age of 24, by claiming to be paranormal on television programs in which he demonstrated his supposed paranormal powers – telekinesis, radiesthesia and telepathy. This illusionist in the mid-1990s appeared on a TV show claiming to be able to transmit symbols telepathically to viewers. He chose one of four symbols: square, star, circle or cross. Millions of people witnessed as he visualized one of the symbols in his mind, urging viewers to tune in to his mental message and phone in their guesses. Almost half of the 70,000 viewers who participated in this viewing chose the star, which he said was being projected "telepathically". In fact, in an excerpt from his book *The Science of Weird Shit*, paranormal psychologist Chris French said that Geller explored a psychological phenomenon known as 'population stereotypes'. Especially in large groups, responses were not equally distributed, but tended to cluster together in a reliable and predictable way. By understanding these patterns, Geller was able to manipulate public perceptions and create the illusion of telepathic powers. Many consider him a charlatan and for this reason, Geller took several people to court who claimed that he did not have paranormal powers, but he lost in all cases.

The CIA analyzed Uri Geller's psychic abilities in 1972 by carrying out tests. With these tests, the CIA concluded their research by saying that he "exhibited paranormal perceptual ability in a striking way". However, after a five-hour on-air meeting with a Brazilian parapsychologist priest, Father Quevedo, famous for his natural explanations of such phenomena, it was demonstrated that what Uri Geller did by bending spoons, forks, knives, keys and breaking watches were just illusionist tricks. The next day, Uri Geller himself discovered that everything was not listening techniques. However, in order not to have to recant in public, Roberto Marinho and Adolfo Bloch, both directors of the Brazilian broadcasters Rede Globo and Rede Manchete, sought an agreement. They promised Father Quevedo that in exchange for an interpretation of Uri Geller he would take the first flight out of Brazil and would never be talked about again on these TV channels [11,12].

8.2.4 INTUITION

Intuition is a person's ability to perceive his or her own subconscious. A lay definition of intuition holds that it involves the immediate grasp of an idea without reasoning. From a more technical point of view, intuition should be viewed as the opposite of logical reasoning, roughly corresponding to the distinction between type 1 (intuitive) and type 2 (reflective) processes in contemporary theories of dual-process thinking. From this perspective, it is known that intuition can provide feelings of confidence, reflect many aspects of information, and is more likely to provide accurate judgments when it is based on relevant experiential learning.

Unlike reasoning, intuition requires little effort and does not compete with the core resources of working memory. It provides standard responses that may or may not involve intervention from high-effort reflective reasoning. Intuition has, however, been blamed for a number of cognitive biases in the psychological literatures on reasoning and decision-making. Evidence indicates that the intervention of intuition in effortful reasoning is often necessary to avoid such biases with new and abstract problems not easily linked to previous experiences. Therefore, although it seems that intuition dominates reasoning most of the time, it can be a false friend, both in the laboratory and in the real world [13].

Intuition is a form of non-conscious, non-verbal and non-deliberate thinking that allows humans to make quick and seemingly instinctive judgments and decisions. Intuition is generally described as a 'gut feeling' or a feeling that something is true without necessarily being able to explain why. Intuition is based on previous experiences and learned patterns, or a form of automation that the human brain uses to quickly and impulsively evaluate situations and problems. However, although instinct can be useful in some cases, it is only sometimes reliable and can sometimes lead to errors in judgment or self-condemnation. For this reason, intuition must often be used with other forms of reasoning, such as analysis and evaluation, to ensure that decisions are more accurately and completely based on concrete facts.

Intuitive decision-making can play a crucial and effective role for business managers. By relying on unconscious knowledge and experiences from the past, managers can make quick, informed decisions even in challenging or uncertain circumstances and attempt further verification later. However, it is essential to also be aware of the possible biases and limitations that may accompany intuitive decision-making. To make the most of their intuition, managers must continually strive to improve their knowledge and experience, reflect on past decisions, solicit feedback, cultivate self-awareness, use data and analytics, and stay well informed. By adopting a multifaceted approach that incorporates intuition, rational thinking and heuristics, decision-making managers can effectively wade through the complexities of their role and make the best, most effective decisions for their organization [14].

From what has seen above, it can be said that intuition is the ability to immediately understand without conscious reasoning (sometimes described as a gut feeling) the correctness or incorrectness of a person, place, situation, temporal episode or object. In contrast, insight is the ability to gain an accurate and in-depth understanding of a problem that is often associated with movement beyond existing paradigms. Examples include the theories of natural selection, relativity, physics, or the unconscious of Darwin, Einstein, Stephen Hawking, Newton and Freud. Many cultures name these concepts and recognize their value, and deeper insight is recognized as a particular characteristic of eminent achievements in the arts, sciences, religions and politics. Considerable data suggest that these concepts are more related to each other than distinct, and that a more distributed intuitive network may fuel a functional neuronal architecture based predominantly on right-sided hemispheric insight.

The preparation and incubation stages of an insight can rely on the incorporation of a domain-specific automated expertise schema associated with intuition. It is known that the neural networks associated with intuition and insight should be better scientifically reviewed. Case studies of anomalous subjects with skill achievement discrepancies are reported in several articles that propose the perspective that the atypical location of cognitive modules can enhance intuitive and insightful functions. This explains individual performance beyond that expected by the skill tests of conventionally measured intelligence. In this branch of knowledge, models and theories of the neuroanatomical basis of intuition and insight are proposed to be used as a starting point for future research to obtain a better understanding of the nature of these two distinctly human and highly complex and still poorly understood abilities [15].

8.2.5 Spells

A spell is an arrangement of objects or scenarios that are believed to be harmful or beneficial to the brain of a person subjected to it. However, sorcerers are attributed supernatural qualities, that is, an irresistible charm that is manifested in such a way as to influence the organization of the target person's brain in such a way as to improve or worsen their life. Therefore, sorcerers can intuitively or experimentally perceive images or arrangements of elements that cause positive or negative influences on other people's brains. The brain of the person who has such power to exercise sorcery can perceive the person being bewitched in terms of their weaknesses, abilities, sadnesses, joys, victories, defeats and what is or is not good or bad for them. These are extrasensory qualities that are not common to all mortals, but that can be manipulated by only a few people.

In *The Tree of Knowledge*, a Jewish potions book, witchcraft and magical spells are seen as items of exposition, as reported in the Hebrew Manuscripts: Journeys of the Written Word. *The Book of Jewish Knowledge (Sefer Hatodaah)*, written by Rabbi Eliyahu Kitov zt"l, presents various phases of the Jewish calendar, such as its days of feasting and fasting, days of joy and sorrow; its laws of observance, in addition to a wealth of Midrash commentaries and inspiring ideas from ancient and contemporary sages [16]. This book was written more than 50 years ago and was published in English several years later, quickly becoming an extremely popular and essential work for everyone in Jewish homes, exploring the breadth and depth of the glorious Jewish tradition and heritage everywhere and at all times.

A small codex from 16th-century Italy titled *The Tree of Knowledge* (*Ets ha-Da'at*) contains about 125 magical spells for all kinds of purposes: curses, healing potions, love charms, woes and amulets. There are a good number of these magical–medical manuscripts in the Hebrew collection, but this publication is special for at least two reasons. First, because of its elegant layout and precision in its execution, and second, because it has an introduction in which Elisha, the author, tells stories, illustrated with several pictures, concerning the way he collected these spells [17].

Sorcery or witchcraft seems to be a subject avoided in higher literature due to having to give greater explanations. This gives the impression that there is a fear of people approaching this subject because it is considered to be of little importance, alongside disbelief, fear, lack of knowledge, shame, or some other special reason. The fact is that this phenomenon is really taken into account, especially in Latin America and Africa.

A sorcerous dispatch can usually be found at crossroads, or *encruza*, which means the intersection of streets in a locality. In Brazil, according to Umbanda, the casting of a spell is a place where offerings are made to Exu and Pomba Gira. Exu is considered a working *Orixá* (in Yoruba: *Òrìṣà*). Exu and Pomba Gira are the deities of the Yoruba religion represented by nature, with Exu being the defender, messenger and guardian of the *terreiros*, villages, cities, houses and *axé*, and of human behavior. Furthermore, it represents communication, patience, order and discipline. The cute Pomba is a spiritual entity known in Umbanda and Candomblé as a female Exu and messenger between the worlds of the *orixás* and the Earth. It is believed that the offerings guided by these entities can have the most varied functions, such as protection, prosperity and relief, among others, all simply called 'dispatch'.

In religions such as Candomblé and Umbanda, offering food at the drop-off is a way of connecting with the *orixás*. These deities were sent by Olodumarê to create the world, and after that teach and help humanity to live on this planet. According to Umbanda, each deity has its corresponding foods for the spell, with its specific manner of preparation and symbolism. These food deliveries also differ depending on the 'strength point' of that entity. For example, Exu, known as the guardian of the paths, has the crossroads as his favorite place, while Iemanjá, who is the queen of the sea, usually receives offerings from the faithful in the waters.

In addition to Umbanda, there is 'lore' that refers to demons at crossroads. For example, a story is told about this in blues music, mainly characteristic of New Orleans. Chicago, on the other hand, has a different style of blues, for example, Robert Johnson is at a crossroads in American music. Rumors circulate that he once stood at a crossroads in Mississippi and sold his soul to the devil in exchange for his unique musical gifts. This all seems to be the strength of the mind, of a strong belief in something that someone really wants and to which the brain submits.

The practice of sacrifices has ancient and magical foundations, representing a dogma for these religions or sects, carried out according to different situations. In Afro-Brazilian religions, animal sacrifice is understood as an exchange of energies between the believer and the immolated animal, when the latter has the purpose of removing negative energies from the adherent, a function known as 'unloading'. There is another type of sacrifice, in which the animal is sacrificed to the Orixá deity as an offering. Each Orixá has a specific animal that they prefer, and this is what is offered to them. This offering, in general, is made once a year during the Orixá festival, which usually also receives other types of offerings, made up of flowers and fruits, just as there are other ways of discharging a person. This way, the sacrifice can be replaced by another practice, if it is not comfortable for the believer. There are, however, situations in which sacrifice is necessary and irreplaceable, in which case it is up to the believer, if this one deems it necessary, to refuse, knowing then that they will not obtain the favors of the *orixá* [18].

8.2.6 MEDIUMSHIP

Mediumship is a person's sensitivity to receiving manifestations from another's subconscious. The word mediumship comes from Latin and means something in between. Since everyone has an electromagnetic field sensitive to a certain frequency of vibration, everyone has the greater or lesser capacity to receive and transmit energies and vibrations, acting on the most diverse planes of existence, from the densest to the most subtle (see Chapters 2 and 3). Believers in this ability think that mediumship is an exchange between heaven and earth, that is, between the spiritual world and the material world. In reality, mediumship is a topic that is widely discussed in one way or another around the world. In Brazil, it was more widely disseminated by the world's best-known medium, Chico Xavier in particular, since this country is considered the homeland of the gospel. He tried to reduce the distance between the invisible and the visible, comforting the hearts of many, spreading his various psychographed works around the world, as is better detailed in Chapter 10 of this book. However, no intermediaries are necessary between the person and spirituality. For some people it may make sense to have a religiosity or spirituality and follow dogmas, but for others it does not, considering that Buddha was not a Buddhist, Jesus was not a Christian, Krishna was not a Vaishnava and Muhammad was not a Muslim. They were simply teachers who taught the forms of sensibilities among human beings [19,20].

It must be said here that mediumship is an innate faculty in all human beings. It is part of each person, therefore, it does not depend on an external factor to develop, that is, it is intrinsic to each being and can manifest itself anywhere as a sixth sense, as the voice of superior consciousness or the soul. It enhances intuition and contact with the higher ego and allows communication directly with the source and each person's guides, mentors and 'guardian angels', without the need for intermediaries. According to Professor Sarah Sofia, everyone is born a medium [17–19]. This is a quality and a right of each person, as everyone is in some way a spiritual being. However, there needs to be merit, sensitivity, ability and a genuine purpose for the full development of mediumship with access to the most subtle fields. The motivation for developing mediumship, first, must be the search for one's own spiritual evolution and, consequently, this will reverberate in other things. Mediumship, as a rule, is a tool in the hands of people who have correct conduct, principles and values. However, in the hands of selfish, manipulative, excessively materialistic people who are only looking to be powerful, advantageous or better than others, it can be a destructive weapon, both for themselves and for others, as this will only attract low vibrations and other problems [20–22].

There are numerous common symptoms reported by people linked to mediumship, which can awaken mediumistic sensitivities [20–22]:

- Easily perceive lies;
- Intuitions and premonitions;
- Feeling emotions about something that appears out of nowhere;
- Have knowledge of things that you do not master;
- See figures and spirits;
- Feel smells that no one else is smelling;
- Feel the effects of the energy of people and environments;
- Have lucid and premonitory dreams;
- Tingling and/or heaviness in the hands due to the desire to write;
- Sensation of touches on the body;
- Know when someone is sick;
- Hearing hums, noises and voices without a defined source;
- Yawning;
- Capture other people's thoughts;
- Tremors or chills;
- Dry mouth;
- Change in body temperature;
- Headaches;
- Tachycardia, sweating and trembling body.

8.2.7 Visions

According to the Britannica Dictionary, a visionary is someone who is having or showing clear ideas about what should happen or be done in the future, having or showing a powerful imagination. Therefore, it does not have much to do with imagination or artistic creations. One interpretation from Wikipedia says that a "vision is something seen in a dream, trance, or religious ecstasy, especially as a supernatural apparition that usually conveys a revelation. Visions are generally clearer than dreams, but traditionally with less psychological connotations. Visions are known to arise from spiritual traditions and can provide a picture for human nature and reality. Prophecy is often associated with visions" [23,24].

The Catholic Dictionary defines a vision as supernatural knowledge in which the mind receives an extraordinary understanding of some revealed truth without the aid of sensible impressions. Mystics describe them as intuitions that leave deep impressions [25]. It is clear from the Bible that dreams and visions are important ways that God chooses to speak to people. This saying is very important to under-stand the language of dreams and visions, so that the person can hear what God is saying at night or during prayer times. Of course, everyone dreams, and nowadays there is great interest in dream interpretation. Interpreting dreams for non-Christians can help them receive God's love and guidance. So, after these explanations, it is very difficult to have a convergent interpretation of any vision [26,27]. Therefore, vision becomes a subconscious montage of the effects of frequencies that generate images in oneself or in other people's brains.

It must be said that both visions and dreams are mental experiences that have important differences despite the fact that they can be significant in teaching someone something. Visions and dreams have different levels of importance and can be interpreted differently in distinct cultures. Both dreams and visions have the ability to offer insights, but visions and dreams often serve different purposes, occurring at different times and having different and unique things to teach a human being.

Although they may seem very similar, the three main differences between visions and dreams can be detailed as follows. Visions come to the person at any time, while dreams only occur while they are sleeping. Visions help to create a connection between the life imagination based perhaps on the person's own culture, the spirits and the Creator. Visions can be important in rites of passage while dreams help people connect to themselves.

Brain injuries are believed to cause a variety of symptoms. Among them, visual disturbances are the most common, as 50–70% of patients experience some change in vision after injury/illness, with one example presented in Chapter 10 [25]. Other very common and disabling symptoms are fatigue, anxiety and depression. Typically, these symptoms are best examined if levels of fatigue, anxiety and depression are increased in any way, at which point patients will also experience visual disturbances. An association was found between visual symptoms and fatigue, but not between visual symptoms and anxiety/depression. However, some visual symptoms, such as glare, blurred vision and reading difficulties, showed large differences between patients with and without anxiety/depression. They may be influenced by what is previously contained within their brains. Vision rehabilitation can be a very useful tool to mitigate fatigue after an acquired brain injury, which can be linked to myopia, anxiety, depression, fears and vision-related quality of life.

The ability to experience mental images is innate in human beings. Therefore, 'mental imagery cultivation' is proposed to identify technological traditions dedicated to the deliberate and repeated induction of enhanced mental imagery, which gener-ally occurs in a few selected individuals. Improving the mental image increases the vividness and control of this mental image due to its functional and adaptive value.

Generally, the cultivation of mental images is embedded in magical–religious traditions and is more or less independent of social complexity. The cultivation of visions in shamanism can be explored as an example. Experimental evidence from the psychological literature is presented that demonstrates the functional equiva-lence of mental imagery and perception at specific non-volitional levels in the psy-chophysiological apparatus, thus suggesting that the shaman experiences 'visions' as 'real' and a reaction at a deep psychophysiological level to their contents. Individual differences in the capacity for mental imagery may be an important determinant of the shaman's social role. Experimental evidence for the functional importance of mental imagery in human memory is presented to suggest that the shaman's legendary superior mnemonic abilities may be due to the development of greater use of mental imagery [26].

In particular, visual impairment includes the population of people known as blind, who cannot perceive any visual stimulus, and people with low vision, visually impaired, and characterized by a variety of symptoms. These symptoms can include tube vision, lack of visual acuity, high sensitivity to light, night blindness and difficulty in distinguishing colors. It seems that visually impaired people see the world by developing other sensory channels, but the senses that most help them in this process of assimilation and understanding are auditory and tactile, contributing greatly to the construction of mental images [27].

8.2.8   PREMONITION

Premonition is an abstract manifestation of the subconscious with an association with facts and knowledge, that is, a kind of intuition to predict a future occurrence, generally undesirable or long awaited. It may also be a premonition or an advance warning of a future event. It is often interpreted as an advance warning of a clairvoyant or clairaudient experience, such as a dream, which resonates with some event in the future because it is a strong intuition that something, usually negative, is about to happen.

A death premonition is a strange and inexplicable feeling that someone is going to die. Some people have a very clear premonition of death, while others just have an overwhelming sense of fear or anxiety about an event or person. A premonition of death does not necessarily mean that the person will die. A premonition may simply be a warning designed to prevent premature death. Not everyone who has a premonition of death is a medium or clairvoyant. This type of premonition can happen to anyone, depending on what has accumulated in this person's brain. People who have a premonition of their own death do not necessarily die; however, people who suffer an accident or other type of trauma often have these premonitions about death.

The National Center for Biotechnology Information conducted a survey of 302 trauma staff about patient premonitions of death (PODs) [28]. These reports expressed a feeling or sensation of death, which results revealed that these patients were actually more likely to die. Other survey results of trauma healthcare professionals included [26]:

- 95% stated that trauma patients expressed fear of personal death (FOP);
- 44% stated that patients could feel the result of their situation;
- 50% stated that PODs have a higher mortality rate;
- 57% believed that patients affected the outcome through the strength of their will/desire to live.

Although vivid premonition is portrayed as a vision, in which a person receives information through which this person is able to see things that will happen in the future, this vision can be much more descriptive than just a presentiment or hunch.

Despite this, it is important to say at this point that not all premonitions of death can be interpreted literally as a real death. This death can be symbolic, such as the end of the current way of life or an unwanted situation, or an exacerbated fear, the death of a career or even the end of a relationship. If the person is on a spiritual path and striving to evolve spiritually, the dream may symbolize their spiritual death and being reborn at a higher level of spirituality [29].

It seems that the existence of predictive dreams can never be proven or disproved by science due to their diversity. Instead, researchers can study larger samples of people who have had precognitive dreams to obtain a broader picture of who has them and how they are caused.

8.2.9 PRECOGNITION

There seems to be a distinction between 'precognition', which is a parapsychological faculty or direct spiritual knowledge of the future, and 'premonition', which is a term from spiritualism referring to warnings or signs that a prudent person will notice and avoid. Precognition is directly associated with premonition, and is considered a supernormal knowledge of future events, with emphasis not on mentally causing the occurrence of certain events, but on predicting those whose occurrence the person claims were already predicted. Like telepathy and clairvoyance, precognition is supposed to act without recourse to the normal senses and is therefore a form of ESP.

There is a long tradition of anecdotal evidence for predicting the future in dreams and in various other ways, such as observing the formation of clouds, the series of events for people, the formation of birds' flight, or something more drastic, such as examining the entrails of sacrificed animals. Precognition was tested on subjects in such a way as to predict the future order of cards in a deck about to be shuffled or to predict the results of dice rolls. However, statistical support for this has generally been less convincing than experiments with telepathy and clairvoyance [30]. Perhaps this can be attributed to a form of subconscious reasoning based on previous experiences.

Precognition can be summarized as a 'lucid estimation' that carries a serious psychiatric association, with some considering that all precognitive people automatically suffer from schizophrenia or worse. Paradoxically, there are more 'lucid estimates' by scientists than by soothsayers and esoteric fans. It is feeling that something, often of an undesirable nature, is about to happen, that is, a prediction of future events related to the ability to predict what will happen or what will be needed in the future [29–32].

8.3   TELEKINESIS, CLAIRVOYANCE AND OTHER DIVINATIONS

Telekinesis, clairvoyance and teleportation are very challenging subjects and very difficult to obtain something concrete on for scientific proof. Not many authors want to get involved in such abstract subjects, sometimes involved with fraud, errors, beliefs or volatile scientific bases. Even so, these themes still have good reasons to attract multiple studies and understanding.

8.3.1 TELEKINESIS

According to Wiktionary, telekinesis is the ability of a person to move objects with the power of their mind. This eternally appealing idea of moving an object remotely using only psychic powers has had a long life in movies, TV shows, stories and novels, video games and comics. *Tele* in Greek means distance and *kinesis* means movement. However, although some researchers believe in the possibility of the existence of telekinesis, also known as psychokinesis, most scientists believe that any reported experiences were the result of fraud, dreams, ignorance, coincidence, or naturally explainable events. Perhaps there could also be a relationship to an event predictable by the telekinetic that previously announces a certain movement.

Although most people do not believe in telekinesis, there are those who feel strongly that other people can move or change objects with their minds. The study of this phenomenon is part of parapsychology, which also includes things such as mind reading and reincarnation. Alexander N. Aksakof, a Russian advisor to the Tsar, first used the term telekinesis in 1890 [33].

8.3.2 CLAIRVOYANCE

Clairvoyance research has produced over decades a significant body of evidence about the mind's ability to collect information not accessible to ordinary physical senses. In the materialist philosophical worldview, mind-to-mind communication is impossible. Therefore, these reports do not happen because they simply cannot happen. However, they continue to occur repeatedly around the world.

In the current creative scientific maelstrom, perhaps it is time to consider the possibility of a model in which clairvoyance, rather than being an anomaly, is a fundamental feature of the Universe. 'Scientific theories' of everything are now emerging that can accommodate anomalous phenomena such as clairvoyance [34].

Ancient publications, such as [35–39], establish that the power of clairvoyance manifests itself in all forms of fact perceptions, happenings and events in relation to future times. Explanations of prophecy, futurology, prediction, a second location, etc. are not supernatural, but are merely the development of clairvoyant faculties. How can something be seen years before it happens? The only reasonable explanation would say that to determine cause and effect there must at least be some form of existence, at least a latent potential for this, or a keen perception of subconscious faculties such as subconscious reasoning. Coming events always cast their shadows long before they can happen. Considering fate versus free will in time is just a relative way of looking at things. Events, in a sense, always exist, both in the past and in the future. Time is like a moving film reel, containing the future scene now, although out of sight. The dreamtime analogy is an absolute consciousness in which past, present and future exist as a simple perception. A glimpse of a transcendental truth is like acquiring the faculty of future clairvoyance [38,39].

8.3.3 TELEPORTATION

Science is still working on an answer to explain teleportation, as it is still very difficult to believe that something or a person can be transported without a physical means.

Maybe it is as simple as scanning a person down to the subatomic level, annihilating all the favorite parts of one place and then sending the scanned data to another distant point. This still has no acceptable scientific basis. At the far point, there is no method or study of how a computer could reconstruct this body from nothing as its complete original parts, all in a fraction of a second or perhaps years [40,41].

8.3.4 TAROLOGY

Tarology is the use of a deck of cards (tarot) to play a 'game' and interpret what appears in the sequence of the cards, giving advice or answering a consultant's questions. Based on this information, the tarot predicts possibilities, even if it does not determine concrete facts. Based on the information perceived by the tarot reader, this deck can interpret the energy of the moment and tell where the person is likely to go. The network of associations of ideas constructed by the tarot has a presumed relationship with the Egyptian, Greek, Christian and Gnostic mystic gods. In some theories, the tarot is identified with the book of the Egyptian god of the moon and wisdom, known as Thoth. This was later compared to the god Hermes. Scientifically, this could be interpreted as the sensitivity of someone who can read desires, sensations and dreams of people with minds sensitized by the positioning of the cards.

8.3.5 CARTOMANCY

A fortuneteller is generally a person who reads a common deck of cards without any specific meaning, but who serves to interpret the set of cards open on the table with the participation of the consultant, who is generally the one who splits the deck. Normally, fortunetellers do not study each of the cards, as they do not contain any specific meaning, but they have learned naturally from experience in predictions using their extrasensory sensitivity that allows them to understand the meaning of the cards displayed. In relation to the tarot reader, the fortuneteller is generally more punctual when representing the past, present and future, being more objective with the questions asked by the consultant. The main difference in relation to tarot readers is that they work with not only the same predictions, but also using a therapeutic approach. From this, the tarot reader explains facts, paths, changes and events that can give guidance on what is presented in the game. Both professionals have the mission of passing on information to the querent, always providing clarification on each prediction addressed regardless of the results of the cards, perhaps suggested by the querent's own mind [42].

8.4 SELF-KNOWLEDGE

Self-knowledge is a very important skill for the emotional balance of people who think they know themselves, but in reality, they do not know themselves that well. Self-knowledge is something deeper, as it helps to understand the origins of people's behaviors over time and their real limits and desires.

   In philosophy, 'self-knowledge' typically refers to knowledge of one's own mental states. That is, what someone is feeling or thinking at that time, or what they believe or desire. At least since Descartes, most philosophers have believed that self-knowledge differs markedly from knowledge of one's external world, which also includes knowledge of other people's mental states. However, there is little consensus on the aspects that could precisely distinguish self-knowledge from knowledge itself in other domains. In part because of this disagreement, philosophers have endorsed competing accounts of how self-knowledge and epistemic status can be achieved. These accounts have important consequences for a wide range of questions in epistemology, philosophy of mind and moral psychology, focusing on knowledge of mental states themselves. A separate topic, sometimes referred to as self-knowledge, is knowledge about a persistent Self.

In psychology, self-knowledge is a term used to describe the information an individual uses when finding answers to personal questions such as "what am I like?", "where am I going?" or "who am I?". When seeking to develop an answer to these questions, self-knowledge requires self-awareness about the emergence of isolated issues and an ongoing self-awareness that should not be confused with pure awareness. Babies and chimpanzees exhibit some of the traits of self-awareness and agency/contingency but are not considered to also have self-awareness [43,44]. At some higher level of cognition, however, a self-conscious component emerges in addition to an increased self-conscious component, and then it becomes possible to ask, "what am I like?". The answer to this is self-knowledge, although this knowledge has its limits, as introspection is considered limited and complex.

Self-knowledge in psychology is real and genuine information that someone has about themselves [45]. This includes information about personal emotional state, personality traits, relationships, behavioral patterns, opinions, beliefs, values, needs, goals, personal experiences, preferences and social identity. Self-knowledge results from self-reflective and social processes. However, self-knowledge is not derived from introspection alone. According to Barron (2002), there are five sources that contribute to the reservoir of self-knowledge [46]. These are: 1) consulting the physical, concrete world; (2) comparing oneself with others, social comparison; (3) incorporating the opinions of others, reflected assessment; (4) looking inside oneself, introspection; and (5) examining one's own behaviors and the context in which they occur, one's perception of oneself and one's attributions [47].

Because self-knowledge includes honest self-assessments and other acquired information, it can be used to make positive changes and master aspects of people's lives. Self-knowledge is essential to "give a meaningful narrative to one's past, present and future actions, and thus give a sense of continuity over time, a sense of being unique and similar to others" [37]. Knowing yourself increases your ability to live a coherent and satisfying life. Furthermore, it allows someone to understand their own basic motivations and fears and improve control of emotions. On the other hand, the inability to recognize one's feelings leaves a person vulnerable and at the mercy of these feelings. Stellar self-knowledge motivates everyone to pursue ambitious projects, new relationships, and other challenges. Lack of self-perception can inhibit high aspirations [47–50].

Research suggests that self-knowledge, or how well one understands oneself, is actually much more challenging than one might imagine. Social pressure, education and psychological defense mechanisms can distort people's perceptions of themselves and lead to a mismatch between how they see themselves and who they really are. Fortunately, with time, effort and a willingness to accept one's flaws, one can come to know one's true self [51].

8.5   BRAIN EVOLUTION

A billion years of evolution appears to have selected five fundamentally different types of brains, each suited to its purpose, scientists say. Andrew Barron and his colleagues at Macquarie University (Sydney-Australia) claim to have identified five major changes in the computational capacity of brains, which allow us to reach the world of known intelligent life. These are a nervous system, a centralized nervous system, a brain with feedback, a brain with multiple recurrent systems and reflection. Even so, there is still a long way to go before it is possible to add a "new type of computational processing", including artificial intelligence [50].

The development of a nervous system to coordinate the actions of the first organisms began in jellyfish because they were equipped with diffuse neural networks that were optimal for coordinating a body. This species can survive massive damage, but its neural networks are very poor at gathering information. The evolution of the first species then led to the development of a centralized nervous system, with a brain capable of acting as the master coordinator and combining information from different senses. For this purpose, leeches and tardigrades that are in this category were studied by the researchers. They then began studying a brain with recurrence, or feedback. Bees were included in this classification because they can quickly learn different types of art and recognize abstract concepts and navigate using brains that incorporate rapid feedback into their actions. A more evolutionary step formed a brain with multiple recurring systems, feeding back information with and between each system. For this, birds, rats and dogs were selected to study because they use massive parallel processing of information, using the same information in several different ways at the same time and recognizing the relationships between different types of information. In addition to these, monkeys can solve problems and create rudimentary tools. With the evolution of these laboratory studies, it was observed that monkeys differ widely in natural evolution, requiring new hypotheses, such as the collective evolution of human intelligence. Then came studies on brain plasticity, as human brains can modify their own computational structure according to what they think is necessary. A reflective brain can learn the best flow of information for a specific task, evolve based on what it has learned and modify the way it processes information in real time to be faster and more efficient.

The human brain is reflective because it enables their imagination, thought processes and the rich mental life they have. Humans also made use of symbolic language, expanding their minds even further by allowing communications between them in a coordinated and very efficient manner. All these types of current studies have led to the development of autonomous machines, yet based on the coordination of jellyfish, the obstinacy of worms, the quick thinking of bees and the complex interactions of birds during movement in their collective flights. It must be noted that bees can do things that human beings simply cannot, as they are fully functional from the moment their wings dry out when leaving their cells of origin. They can learn to navigate for miles around their hives without losing their way back. Different types of brains adapt all animals to different lifestyles, including humans, which is why they can still share the same planet with jellyfish and worms essentially unchanged for hundreds of millions of years. The brains of these beings are simply perfect for what they need to do. Much of this ability of worms can help in creating new types of intelligence for autonomous machines and artificial intelligences [51–53].

Human transitions are cumulative and not progressive. The accumulation of knowledge depends on where it came before, but without evolution in relation to what came before. Brains are part of embedded systems that are adapted to different niches and they are electrochemically connected to the body. A multicellular organism has lost certain possibilities that remain open to single-celled creatures. Just as the shift from single-cell to multi-cell life requires individual cells to work together in the same direction rather than optimizing their individual destinies.

New ways of organizing the brain bring new demands, new limits and new opportunities. Fuzzy neural networks are intrinsically parallel, for example, while the centralization that a network produces with a serial bottleneck of subsequent transitions only incompletely overcomes itself. Centralized architectures may work well out of the box, whereas recurrent architectures require much more training to be useful, in addition to the fact that a large number of organisms get along perfectly well without any brain or nervous system. The benefits of transitions are that they help science understand the big picture of cognitive evolution and structure the diversity of animal minds.

The mappings of human transitions are different in kind and not just in positioning. Cognitive transitions are not just about obtaining answers to what transitions have occurred, but to what a transition means in the first place. This fact places changes in computational architecture as strong, compelling candidates for major transitions in cognitive evolution and shows that common resource constraints can alter computational architectures for a radically different transformation of subsequent evolution.

The focus on computational architecture has the generality needed to capture large-scale patterns in the life history. Previous reports have focused on specific cognitive abilities or methods of adaptation and learning. However, these capabilities depend on an even broader computational architecture, including learning types with a space of possible algorithms that can be explored, while the computational architecture determines the space itself [54,55].

Consciousness is one of the last biological phenomena about which human beings still do not have a clear idea to answer how and when it emerged and evolved. It is important here to know how to identify the adaptive value of consciousness, the relationships between the brain, behavior and consciousness that must be understood. Thus, the evolution of the mind is closely linked to the question of how it comes from the brain through the mind–body relationship. It seems that the evolution of consciousness cannot be solved without first solving the 'hard problem' of why the evolution of consciousness based on the evolution of cognition is premature and unfalsifiable [56,57].

There are publications exploring some established facts in neuroscience to analyze the information processing by a population of neurons. Most of these publications have adopted nonlinear analysis methods with neuroscience as a basis for understanding the intrinsic nature of conscious information processing. They aimed to gain some kind of mastery over the elusive concept of phenomenal experience. In these analyses over the years, neuroscientists have employed various imaging techniques to discover a multitude of new and complex facts that have uncovered many peculiarities of the central nervous system. Analytical methods developed in nonlinear sciences and information theory, among others, can be used to take advantage of enormous knowledge and thus improve the structures currently suggested [57,58].

Like everything that happens in the evolution of species, the brains of animals also evolved into what they have become today, as those who did not adapt died and thus only those who were truly capable of surviving the current conditions of their time were left. Obviously, living beings will become increasingly adapted to the natural conditions existing in the environment in their times, relating to air, water, food, the form of reproduction, the available space and natural kinetic and static energies, among others. The naturalist Charles Darwin (1809–1882) established the basic knowledge of how natural selection happened and how the recurring competition for the survival of each species selected the one that was best adapted to the environment where it was generated. Those who did not adapt simply died and disappeared. The result of this seems somewhat miraculous to current humans impressed by the apparent 'perfection' of the current constitution of living beings after so many billions of years. Simply, these beings gradually adapted to temperature, pressure, humidity, radiations, relationships, the environment and everything else to which they were exposed over billions of years. Fish adapted to the sea, humans adapted to land, birds adapted to flying in the air, bacteria to living in living bodies, and so on. The same apparent 'miracle' will possibly happen in another billion years of evolution.

REFERENCES

[1] American Psychological Association, Publication Manual of the American Psychological Association (APA), 7th ed., https://doi.org/10.1037/0000165-000, 2020.

[2] C. Small, The Science of Mind Control and Telepathy, Unabridged ed. Cavendish Square Publishing, ISBN-10: 1502637995, ISBN-13: 978-1502637994, 30 Jul 2018.

[3] G. Venkatasubramanian, Investigating paranormal phenomena: Functional brain imaging of telepathy (Peruvumba N. Jayakumar, Hongasandra R. Nagendra, Dindagur Nagaraja, R. Deeptha, and Bangalore N. Gangadhar), International Journal of Yoga, 1(2), 66–71. https://doi.org/10.4103/0973-6131.43543, PMCID: PMC3144613, PMID: 21829287, Jul–Dec 2008.

[4] M.A. Persinger and F. Healey, Experimental facilitation of the sensed presence: Possible intercalation between the hemispheres induced by complex magnetic fields, The Journal of Nervous and Mental Disease, 190, 533–541, 2002, PubMed. Google Scholar.

[5] P.L. Jackson, E. Brunet, A.N. Meltzoff, and J. Decety, Empathy examined through the neural mechanisms involved in imagining how I feel versus how you feel pain, Neuropsychologia, 44, 752–761, 2006, PubMed. Google Scholar.

[6] J. Pascual-Leone, Reflections on working memory: Are the two models complementary? Journal of Experimental Child Psychology, 77(2), 138–154. https://doi.org/10.1006/jecp.2000.2593, PMID: 11017722, Oct 2000.

[7] K. Cao, M. Ma, C. Wang, J. Iqbal, J. Si, Y. Xue, and J. Yang, TMS-EEG: An Emerging Tool to Study the Neurophysiologic Biomarkers of Psychiatric Disorders. National Library of Medicine, National Center for Biotechnology Information, https://doi.org/10.1016/j.neuropharm.2021.108574, PMID: 33894219, 01 Oct 2021.

[8] A. Mantovani, M. Aly, Y. Dagan, A. Allart, S.H. Lisanby, Randomized sham controlled trial of repetitive transcranial magnetic stimulation to the dorsolateral prefrontal cortex for the treatment of panic disorder with comorbid major depression, Journal of Affective Disorders, 144(1–2), 153–159. https://doi.org/10.1016/j.jad.2012.05.038, 2013.

[9] A. Datta, V. Bansal, J. Diaz, J. Patel, D. Reato, and M. Bikson, Gyri-precise head model of transcranial direct current stimulation: Improved spatial focality using a ring electrode versus conventional rectangular pad, Science Direct, Brain Stimulation. Elsevier, 2(4), 201–207. https://doi.org/10.1016/j.brs.2009.03.005, Oct 2009.

[10] H.S. del Moral, La rabdomancia en la Nueva España. Práctica, apología y ridiculización, Relaciones. Estudios de historia y sociedad, Dowsing in New Spain, Practice, Apology and Ridicule, 40(160), https://doi.org/10.24901/rehs.v40i160.602, ISSN 2448-7554 (On-line), ISSN 0185-3929 (press), Zamora dic. 2019, Epub, 19 Nov 2020.

[11] J.O. Green, Uri Geller and the reception of parapsychology in the 1970s, Faculty of Graduate and Postdoctoral Studies, Master of Arts in History, The University of Chicago, Jul 2018.

[12] Nature: Uri Geller, Vol. 251 Oct 18 1974, CIA-RDP 96-0078R000200090024-7 Parapsychology Foundation, Approved Release: 10 Aug 2000.

[13] St B.T.E. Jonathan, Intuition and reasoning: A dual-process perspective, Special Issue on Intuition (Taylor & Francis), 21(4), 313–326, www.jstor.org/stable/25767 204, Oct–Dec 2010.

[14] A. Bensla, Intuitive decision making: Its' Pros and Cons and 4 models, Risely, www.risely.me/intuitive-decision-making-pros-and-cons-and-4-models/, 15 Feb 2023.

[15] S.M. McCrea, Intuition, insight, and the right hemisphere: Emergence of higher sociocognitive functions, Psychology Research and Behavior Management, 3, 1–39. https://doi.org/10.2147/prbm.s7935, PMCID: PMC3218761, PMID: 22110327, 03 Mar 2010.

[16] P. Baroja, The Tree of Knowledge (El árbol de la ciencia), (English Edition), eBook Kindle, original from Spain, 1911.

[17] Elisha ben Gad of Ancona, The Tree of Knowledge: Magic Spells from a Jewish Potion. British Library, Asian and African studies blog, https://blogs.bl.uk/asian-and-african/2020/08/the-tree-of-knowledge.html, 19 Aug 2020.

[18] M.F.P. Amorim, Sacrifícios rituais em religiões afro-brasileiras: A proteção jurídica aos animais não humanos frente a valores religiosos e culturais (Ritual sacrifices in Afro-Brazilian religions: Legal protection for non-human animals in the face of religious and cultural values), Magazine Jus Navigandi, 19(4082), ISSN 1518-4862, Teresina-Br. Available in: https://jus.com.br/artigos/31559, 04 Sep 2014.

[19] G. Lucchetti, H.G. Koenig, and A.L.G. Lucchetti, Spirituality, religiousness, and mental health: A review of the current scientific evidence, World Journal of Clinical Cases (Baishideng Publishing Group), 9(26), 7620–7631, https://doi.org/10.12998/wjcc.v9.i26.7620, PMID: 34621814, PMCID: PMC8462234, Sep 2021.

[20] S. Sofia, Mediunidade: sintomas, tipos e como desenvolver, https://guiadaalma.com.br/mediunidade-sintomas/, 14 Jul 2022.

[21] S.B. Milate, 9 Sintomas de Mediunidade de Incorporação: Iniciante e Involuntária, Luz dos Anjos, Blog de Espiritualidade, www.luzdosanjos.com/blog/sintomas-de-mediunidade-de-incorporacao/, 05 Aug 2021.

[22] N. Cardoso, Sintomas da Mediunidade, Portal do Espírito, https://espirito.org.br/arti gos/sintomas-da-mediunidade/, 04 Feb 2019.

[23] D.A. Schreuder, Vision and Visual Perception. Archway Publishing. ISBN 978-1-4808-1294-9. Retrieved 13 Jul 2018, 2014, p. 671.

[24] C. Fillmore, The Interpretation of Visions, Unity School of Christianity, www.truthun ity.net/tracts/charles-fillmore-the-interpretation-of-visions. Accessed 29 Jul 2023.

[25] Intellectual Vision, catholicculture.org, Fr. John Hardon's Modern Catholic Dictionary, Eternal Life. Retrieved from www.catholicculture.org/culture/library/dic tionary/index.cfm?id=34262, 16 Jul 2018.

[26] E. Evans, Interpreting Dreams and Visions, A Practical Guide for Using Them Powerfully to Impact the World. Monarch Books, ISBN-13: 9780857217790, 9780857217806, 19 Jan 2018, p. 416.

[27] P.F. Stone, How to Interpret Dreams and Visions, ISBN 978-1-61638-350-3, BR115. D74S76 2011 248.2'9-dc22, ISBN: 978-1-61638-426-5, 2011.

[28] M.A. Miglietta, G.I. Toma, S. Docimo, R. Neely, A. Bakoulis, E. Kreismann, Premonition of death in trauma: a survey of healthcare providers. American Surgeon; 75(12), pp 1220-1226. PMID: 19999916. Dec 2009.

[29] S. Painter, Exploring Premonitions of Death and Their Meanings, Health and Wellness, www.lovetoknow.com/life/grief-loss/exploring-premonitions-death-their-meanings, 30 Oct 2020.

[30] National Institute of Neurological Disorders and Stroke, Brain Basics: Understanding Sleep, www.ninds.nih.gov/health-information/public-education/brain-basics/brain-basics-understanding-sleep, 13 Aug 2019.

[31] K. Fukuda, Most Experiences of Precognitive Dream Could be Regarded as a Subtype of Déjà-vu Experiences, Sleep and Hypnosis, pp 111–114, 4:3, 2002.

[32] M. Valášek, C. Watt, J. Hutton, R. Neill, R. Nuttall, and G. Renwick, Testing the implicit processing hypothesis of precognitive dream experience, Consciousness and Cognition, 28, 113–125. https://pubmed.ncbi.nlm.nih.gov/25062119/, 2014.

[33] J. Summer and A. Singh, What Are Precognitive (Premonition) Dreams? Sleep Foundation, an Onecare Media Co., 31 Jan 2023.

[34] The Editors of Encyclopaedia. Psychokinesis, Encyclopedia Britannica, www.bri tannica.com/topic/psychokinesis. 16 Oct 2018.

[35] S. Panchadasi, A Course of Advanced Lesson in Clairvoyance and Occult Powers, Advanced Thought Publication Co., Health Research Books, first edition 1916, Kessinger Publishing, 01 Apr 1996.

[36] J. Franklin, A Practical Investigation into the Truth of Clairvoyance, Schultze and Co., 1854.

[37] S. Panchadasi (original name William Walker Atkinson), A Course of Advanced Lessons in Clairvoyance and Occult Powers, Book, Advanced Thought Publications Co., Chicago Ill., 319 pp, 1916.

[38] Z. Weaver and K. Janoszka, The Mind at Large: Clairvoyance, Psychics, Police and Life after Death: A Polish Perspective, White Crow Books, ISBN: 978-1-78677-212-1, 07 Feb 2023.

[39] J.G. Taylor, Science and the Supernatural: An Investigation of Paranormal Phenomena Including Psychic Healing, Clairvoyance, Telepathy, and Precognition by a Physicist and Mathematician. Temple Smith. ISBN 0-85117-191-5, 1980, p. 83.

[40] A. Morey, Teleportation, Blastoff! Discovery, ISBN-10: 1644872641, ISBN-13: 978-1644872642, 01 Ago 2020.

[41] K. Bonsor and Robert L., How Teleportation Will Work, https://science.howstuffwo rks.com/science-vs-myth/everyday-myths/teleportation.htm. Accessed 01 Aug 2023.

[42] C.R. Wilson and M. Neuhaus, What Is Self-Knowledge in Psychology? 8 Examples & Theories, Positive Psychology, https://positivepsychology.com/self-knowledge/, 22 Jul 2021.

[43] A. Morin and F. Racy, Dynamic self-processes, in The Handbook of Personality Dynamics and Processes. Academic Press, 2021, pp. 365–386.

[44] T.D. Wilson, and E.W. Dunn, Self-knowledge: Its limits, values, and potential for improvement, Annual Review of Psychology, 55, 493–518, 2004.

[45] A. Morin and F. Racy, Dynamic self-processes, in J. F. Rauthmann (ed.), The Handbook of Personality Dynamics and Processes. Elsevier, 2021, pp. 336–386.

[46] A.B. Barron, M. Halina and C. Klein, Transitions in cognitive evolution, Proceedings of the Royal Society, Series B: Biological Sciences, vol.: 290. Doi: 10.1098/rspb.2023.0671, 2002.

[47] H. Bukowski, Self-knowledge, in V. Zeigler-Hill and T. K. Shackelford (eds.), Encyclopedia of Personality and Individual Differences. Springer, 2019, pp. 61–76.

[48] A.K. Schaffner, What's So Great about Self-knowledge? Psychology Today. Retrieved 31 May 2021, from www.psychologytoday.com/us/blog/the-art-self-improvement/202005/whats-so-great-about-self-knowledge, 25 May 2020.

[49] S. Begley, How much self-knowledge is too much? Mindful. Retrieved 09 Jun 2021 from www.mindful.org/how-much-self-knowledge-is-too-much/, 18 May 2020.

[50] T.D. Wilson and E.W. Dunn, Self-knowledge: Its limits, value, and potential for improvement, Annual Review of Psychology, 55(1), 493–518, 2004.

[51] A.B. Barron, M. Halina, and C. Klein, Transitions in cognitive evolution, Proceedings of the Royal Society, Series B: Biological Sciences, 290. https://doi.org/10.1098/rspb.2023.0671, 2002.

[52] G. Roth and U. Dicke, Evolution of the brain and intelligence, Trends in Cognitive Sciences, 9(5), 250–257, May 2005.

[53] M. Tomasello, J. Call, and B. Hare, Chimpanzees understand psychological states – the question is which ones and to what extent, Trends in Cognitive Sciences, 7(4), 153–156. https://doi.org/10.1016/S1364-6613(03)00035-4, Apr 2003.

[54] L. Radinsky, Primate brain evolution: Comparative studies of brains of living mammal species reveal major trends in the evolutionary development of primate brains, and analysis of endocasts from fossil primate braincases suggests when these specializations occurred, American Scientist, 63(6), 656–663. JSTOR, www.jstor.org/stable/27845782, 1975.

[55] S. Ginsburg and E. Jablonka, The evolution of associative learning: A factor in the Cambrian explosion, Journal Theoretical Biology, 266, 11–20. https://doi.org/10.1016/j.jtbi.2010.06.017, 2010.

[56] W.B. Kristan, Neuronal decision-making circuits, Current Biology, 18, R928–R932. https://doi.org/10.1016/j.cub.2008.07.081, 2008.

[57] Y. Gutfreund, The mind-evolution problem: The difficulty of fitting consciousness in an evolutionary framework, Frontiers in Psychology, Sec. Consciousness Research, 9, https://doi.org/10.3389/fpsyg.2018.01537, 24 Aug 2018.

[58] D.J. Chalmers, How can we construct a science of consciousness? Annals of the N.Y. Academy of Sciences, 1303, 25–35. https://doi.org/10.1111/nyas.12166, PMID: 24236862, 2013.

9 Unconscious Information Processing

9.1 INTRODUCTION

Processing information in the brain begins with stimuli generated by sensory organs such as the nose, eyes, smell, taste and the entire area of the skin. These stimuli provide information such as pain, pressure, smell, taste, fears and colors, among others. For each piece of information reaching the brain through neuronal communication, there is a brain response after voluntary processing depending on each person's will, or involuntary processing depending on the parasympathetic nervous system. Of note, all the brain information can be acquired since the mother's womb and can be recovered by some stimulus coming from the body's sensors. Although the human body changes its cells completely to other cells over a period of approximately seven years, the neurons remain, and the number at birth is the maximum. These neurons can degenerate like cells, compensated, however, by the brain's ability to reconfigure itself with neurons that continue to be alive and operational, which is absolutely phenomenal. For example, imagine that the person goes out for a walk and immediately sees someone walking (stimulus), recognizing their appearance (processing) and then waves and says "Hello, good afternoon" (response). This is typical of living beings as discussed below.

9.2 INFORMATION PROCESSING

Sometimes, life gives a person the opportunity to be proud of getting the result of an expression like: "Didn't I say so?" or "I told you so!" or "That's me" and so on. There are those who believe that they are actually very good at predicting the future. Nevertheless, in most cases, the person only knows that something is going to happen after it has already happened, as it was just one of the possibilities to be considered. The idea of predicting the future has been around since ancient times.

The ancient Chinese had the I Ching to predict the future, while Greek oracles preferred to look for answers in the entrails of animals [1]. Today, intelligence agencies around the world mainly rely on the opinion of experts who have a great deal of information to predict events. Nevertheless, there are ordinary people who routinely outperform experts when it comes to making accurate predictions about the future. They are called superforecasters and, even though it may seem like charlatanism,

DOI: 10.1201/9781003604037-9

there is no fraud involved. This does refer to some type of psychic seer or something similar, emphasizes David Robson, author of *The Intelligence Trap* [2]. Instead, scientists have discovered certain personality traits and specific abilities linked to some people whose intelligences are atypical. There are people who can predict, for example, whether a civil war will break out in a troubled region or who will win the Olympics, Robson told BBC Crowd Science in June 2021 [3]. These people have a natural talent for examining evidence and seeing where the future will lead.

The US Central Intelligence Agency (CIA) and the Federal Bureau of Investigation (FBI) have both had teams of 'psychics' for many decades who are also consulted in specific operations [4]. An interesting case is that of a team led by political scientist Philip E. Tetlock, who invited thousands of participants from all occupations to test their abilities to predict future events starting in 2011 – this spanned four years, 500 questions and more than a million predictions. After this survey, the 2% most successful were called superforecasters [5]. The project later grew into a commercial forecasting company run by Tetlock, whose previous work had shown that professional forecasters were actually not very accurate. After analyzing 82,361 predictions made by 284 experts in fields such as political science, economics and journalism, he concluded that chimpanzees throwing darts at possible outcomes would likely get similar results, as he made clear in his book *Expert Political Judgment* (2005) [6]. "Could these superforecasters, who were not considered experts, play a better role in forecasting?" this political scientist wondered.

As neurons emit chemical and electrical signals to carry out synaptic communications, which are impulse currents as in an electrical circuit, they can be easily affected by electromagnetic radiation from nature and from the current high-frequency communications environments in which human beings live. The chemical signal becomes involved in short-distance communication between nearby cells, while the electrical current pulses emitted by neurons are in long-distance communication. The long distance for the chemical signal can be considered as the journey it needs to take from the brain to the muscles, resulting in the need for the person to move, blink their eyes, move a toe on their left foot or walk faster.

It is very difficult presently to know what is true or false in ideas coming from stories, religions, beliefs, ignorance ... today's world is surreal really, where everything is based on impressions from the senses, excess information or intentions that do not always have the same effects for everyone.

9.3   REPORTED CASES OF PREDICTIONS AND EXPLANATIONS

Futuristic predictions have been very common throughout history, reporting people with the ability to anticipate a vision of events. These people were often curious, open-minded and willing to seek evidence and question their assumptions, thus becoming susceptible to this information. They were probably also intellectually humble, so they were able to recognize their own prejudices and consider them, says David Robson. It was not just about hearing or reading many opinions, but having the ability to update predictions or opinions based on the information found ... and, by the way, not everyone can do that, because they are often very connected to their own beliefs and social environment.

Supermeteorologists are very good at simply abandoning what they thought was correct and adopting another opinion. They are psychologically distinct, Tetlock told the BBC in 2015. Although most people think, of their beliefs, education as something very precious, if they had to identify something in particular or something even being sacred, superforecasters tend to see their beliefs as hypotheses to be tested, which must be revised according to new evidence. This means they tend to be better at making initial estimates as soon as they are asked a question, but they are even better at updating what they think as they get more information, so they can recalibrate if the probability is higher or lower, this political scientist explained [7].

An example of beliefs as hypotheses to be tested is the estimation work of David Hawkins and Stan Ulam who investigated the branching problem in the multiplication of neutrons in a nuclear chain reaction. Stan Frankel and Richard Feynman approached this same problem using classical physics. Nevertheless, Hawkins and Ulam approached it using probability theory, creating a new subfield now known as branching process theory. They investigated branched chains using a characteristic function. After the war, Ulam would expand and generalize this work by describing Hawkins as the most talented amateur mathematician he knew [8].

Norbert Wiener (1894–1964) was an American mathematician, known as the founder of cybernetics. He, in turn, realized that the information that happens in the world reaches living beings through the eyes, ears, skin and other sensory receptors, as these function as instruments that select only some of the information received. Otherwise, such information would overload the mind. This information can also be studied statistically, regardless of the meaning it may have. For example, observing the frequency with which certain symbols occur can break several types of codes [9]. Wiener died before the microcomputer revolution took place, but he predicted and wrote about many of the problems that would arise in this new technology.

The power of intuitive decision-making can be especially important when a person is processing a large amount of complex information that is difficult to remember accurately in many details. In these cases, there may be some benefit in letting the brain wander to another unrelated activity while the unconscious mind sifts through the information and makes the most appropriate decision for itself.

In a series of experiments, researchers presented participants with lengthy details about a series of apartments for purchase. After forming their first impressions, some of the participants were encouraged to weigh the different options consciously before making their decisions. The others were asked to solve a series of word games, a distraction designed to prevent participants from using analytical processing to reach a final decision about the apartments. Surprisingly, the researchers concluded that participants who had thought more carefully about their choices had considerably more difficulty choosing the apartment that, objectively, had the most attractive attributes. Their attempts to analyze the different options had impaired their judgment, leading them to choose one of the less desirable options. However, people who had become distracted by the word games were forced to rely on their intuitive impressions that turned out to be more accurate [10].

Although some studies have suggested that one can proceed immediately with the first impression, there often appears to be an advantage in delaying the decision while taking time on some other distant activity. According to Marlène Abadie, a cognitive psychologist at the University of Aix-Marseille, in the south of France, this pause allows the unconscious mind to create a more accurate summary of the complex information that has been presented. She noted that, in turn, this pause increases the accuracy of intuitive judgment. This guidance can be useful in many similar scenarios, where a person is forming their impressions after general information overload, she said. It may be relevant whenever the person needs to choose between different consumer products described by different attributes, such as cell phones, computers, producers, televisions, sofas, refrigerators or stoves. When purchasing these products, this person may prefer to breathe a little, get some air, have a coffee, or leaf through a magazine, for example, before making their final decision [10].

9.4 SOME EXPLANATIONS FOR PREDICTIONS

There is an important technique, especially for people who suffer from anxiety, in interacting with others, in meetings, or in public dialogues. Generally, these people are advised not to respond immediately, but to pause in their own head, count from one to ten, look at everyone and then respond. This small silence is very powerful, because in a world as active as the one human beings live in, an intentional pause makes people pay attention to the details of their actions. It can be said that there is always time to think. In the meantime, the individuals can then create a question or a more empathetic response that may be better received, making the current information overload more under their control.

Joseph Murphy wrote 36 books on applied science, the most famous of which is *The Power of the Subconscious*, which has been translated into 17 languages worldwide. In this work, he states that the subconscious mind, upon accepting an idea, immediately begins to put it into practice. Therefore, the only thing that is necessary is to get the subconscious mind to accept the idea. With this, the very law that governs the subconscious will bring health, tranquility or the new desired position. However, the subconscious mind accepts everything that is impressed upon it, even if it is false, and will try to bring about the results that must necessarily follow, because the person consciously accepted the fact as true. Therefore, he suggests that people use autosuggestion in the moments before sleeping, when the conscious mind is passive and will not resist the idea that one wants to impress on the subconscious mind [11,12].

The energy of consciousness depends on the patterns of feelings, external influences, thoughts and conduct of each person. Among the most common phenomena generated by this type of consciousness energies are the colors of the aura. The aura is a field of a still little known nature, with some electromagnetic characteristics, with a luminous appearance for sensitive people. Its colors are probably linked to the field energy, the person's health, their thinking about the world and the activities and thoughts of whatever is being surrounded by the aura. For example, living beings, men, women, children, animals, plants, minerals and physical objects, all have some aura with their own characteristics, as discussed in more detail in Chapter 2.

9.5   INTERPRETATION OF AURA COLORS

The aura is a composition of vibrations that emanate from living beings and plants. In vibrational terms, each person's aura is nothing more than frequencies according to the laws of electromagnetism that can generate in certain people a feeling of attraction associated with a feeling of well-being and serenity. They can also generate an unpleasant sensation depending on the interaction of this energy in relation to a person to whom these people are physically close. These electro-thermal physical phenomena manifest the power that the aura has in terms of communication and are related to the need to maintain a sphere of good vibrations in thoughts in order for these to be reflected in each person's emanations [13].

In any person, there are beliefs that form a unity between the general state of the body, mind and soul and the vibratory field around them. It is said that states of tension and illnesses become visible in the colors of the aura, even before they manifest themselves physically. Many people have this ability or defect. It all depends on how each person wants to qualify it. Tables 9.1 and 9.2 list some of the interpretations of aura colors and their meanings that can be found in non-scientific publications based on examples of people who may or may not have such sensitivity or who are dedicated to this topic [14]. Exercises are proclaimed so that people can 'see the aura' through their sensitivity to more colors than the usual range of frequencies. Several esoteric societies and some self-help authors describe this, for example, Lobsang Rampa's book, dealing with how average normal human beings can train themselves to 'see an aura'.

TABLE 9.1
Aura colors and their meanings (alternative 1)

Aura color	Meaning
Violet	Heightened intuition and activation of psychic powers and the awakening of extrasensory gifts
Blue	High capacity for verbal and visual communication and presence of marked creativity
Yellow	Intelligence, high reasoning capacity, lucidity and a lot of optimism
Red	Excessive worries about money, obsessions, anxiety or high nervousness
Black	Very negative feelings such as anger, hatred and the intention to control and dominate others
Indigo	Intuition and deep sensitivity. It is the color of the Aura of Indigo children (dark blue auras)
Green	Balance of emotions and thoughts. Shows a strong love for nature and animals
Orange	Vitality and good health, but it can be linked to everyday stress and addictions
Light pink	Love, tenderness, sensitivity, art, affection, purity and compassion
Gray	Depression, sadness or discouragement with life

Source: [14].

TABLE 9.2
Aura colors and their meanings (alternative 2)

Aura color	Meaning
Violet	Ability to transform suffering into something positive and closer to emotional, psychic and spiritual balance
Blue	Healing through one's own spiritual and mental energies
Yellow	Ability to give, receive and perform group work
Red	Lots of courage, vitality, excitement and strong sexual energy
Green	Health and vigor, happy people in life
Silver	Natural medium with healing capacity, simply by working on self-knowledge and understanding your spiritual gifts
Crystal	With bright and white mist, usually present in the hands of people who heal and who have telepathic gifts, mediumship, purity and kindness

Source: [15].

It is important to point out naturally at this point that one person's human body is not mathematically the same as anyone else's. In other words, if one person can see colors from red to violet, another person may have a neural or physical characteristic that can go far beyond red (infrared colors) or beyond violet (colors beyond ultraviolet), or they may not even be able to see either the color red or violet. A typical case of this limitation are colorblind people who, as a rule, can only perceive white, black and gray. However, certain women known as tetrachromics can see more intermediate colors than men can, as discussed in Chapter 2. Therefore, aura vision is nothing more than something uncommon among people who have such sensitivity that it can be qualified as visual defect or ability to see colors beyond the usual, that is, beyond infrared or ultraviolet colors.

An interesting phenomenon is visual changes with age. This can make people susceptible to beliefs in accordance with this visual change, generally related to the things they learned in their education, the type of literature they study, the way of life they have, their health, religion, beliefs, personal problems and family creation. Many modern buildings consider age when deciding on building colors, both internal and external [16]. Table 9.3 shows the relativity of the tone with which we see the world according to the evolution of human beings' age.

According to some authors such as Jacob Olesen, there are different meanings for the aura colors in relation to the human body [18–20], more or less as follows.

1. **Red**

    This color is associated with feelings of passion and sexual desires. When this color appears in a person's aura, it means that they are grounded in their life goals. A person with red in their aura is emotionally and psychically grounded or balanced. Wealth and material expenditures are a fun game, and these people hate denying themselves the simple pleasures of life. Red symbolizes the zest for life.

TABLE 9.3
Variation in the tone of the world with age

Age	Tone of the world
Newborns	They see very poorly, seeing the world blurry and are still unable to direct their eyes to a fixed place
2 months	They have seen in colors since they were born, but it is only around this age that they can distinguish similar tones, such as red and orange
4 months	They can identify faces and babies develop depth perception, realizing whether something is close or far away
8 months	Vision is complete, but unstable, as the eye continues to grow, which can cause disorders throughout life, such as astigmatism and myopia
8 years	Neural connections are stabilized, learned visual functions are no longer forgotten and the world is seen in a bluer tone
23 years	At this age, men's eyes stop growing and women's eyes stop growing at 25, which is why myopia surgery is already suggested at this age
40 years	From this time on, the lens starts to lose its elasticity and difficulties arise in seeing up close
60 years	The lens begins to become opaque and yellowish, causing cataracts and the world begins to be seen in more reddish and yellowish tones

Sources: [16, 17].

2. **Orange**

An orange aura signifies someone's happiness with his or her friends, family and environment. A person with much orange in their aura is quick to make and keep friends. Orange is the color of the Sacral Chakra, which refers to the person who will keep their negative or positive emotions as they are influenced by the relationships they maintain with others.

3. **Yellow**

A yellow aura signifies the inner happiness and balance that these people have within themselves. Therefore, it is the color of the Solar Plexus chakra. Yellow can also indicate a playful spirit, high self-esteem, a spiritual awakening, a heightened intellect or a pang of hunger for greatness.

4. **Green**

It will be a surprise for many people that green is the color of self-love and not the color pink, although both have about the same frequency levels. When green is present in a person's aura, it means two things: either they are in love with someone who balances them, or they have a kind and loving heart. In other words, a loving kindness towards animals, plants, friends, family and life in general, to put it more precisely.

5. **Blue**

Blue is the color of communication; therefore, it is the color of the throat chakra. A blue aura reveals someone who enjoys meditation, is in a calm state and protects those they care about. They are often a support system for friends and family.

6. **Indigo**

   The color indigo is known for allowing a person to see other people's energies. The color indigo in an aura indicates someone who is in tune with their higher self-meaning, one who seeks truths from the unknown and who can feel the energies of other people. It also reveals a power that is used to see beyond the deception that people try to pass off as truth.

7. **Purple**

   Purple is the highest level of all chakra colors. As the color of the crown chakra, people who have purple in their aura are intuitive and big-picture seekers, loving to guide others to their highest potential. They are often the artistic type.

8. **Pink**

   As mentioned above, both the pink and green colors vibrate at the same frequency level. However, a pink aura shows someone who is happy and in harmony with those around them. Someone who has this aura can often be kind, whether to others or to themselves.

9. **Silver**

   Silver is the color of abundance, which can mean that a person is acquiring spiritual or material wealth.

10. **Brown**

    This color should be considered a warning to some, as brown signifies emotions of greed and selfishness.

11. **Black**

    Although it is not a bad color in the aura when black is present, it means that the person has a large amount of anger or sadness built up within them. It also means that they have not forgiven what happened to them and are still holding on to that pain. This anger and suffering arising from past or recent events may be directed at themselves or at other people. It can also symbolize health problems.

12. **White**

    White is the color of energetic protection and, when it is present in an aura, it means two things: the first is that the person is more concerned with spiritual matters done on this earthly plane and, second, that they do not care as much about material goods or needs. It can also symbolize a healthy individual.

9.6 HOW BRAIN MANIFESTATIONS CAN HAPPEN

Information in the human brain is received through the body's sensors and can come in the form of smell, temperature, natural color, infrared color, ultraviolet color, sound, ultrasound, infrasound, tone of voice, previous experiences, vibrations at the most different frequencies and others. In short, it is an immense arsenal contained in the human being's subconscious as it has been accumulated throughout their life, including the gestational period. This information is stored in the side of the brain known as the id, in an illustration of visual proportions shown in Figure 7.2 of Chapter 7 in relation to the ego and the superego. Such information can be revealed at any time when there is a provocation.

The brain perceives throughout life through the senses all the dangers and needs that affect it according to what it has already accumulated. These frequencies have been recorded in one way or another in the neurons in such a way that every time there is a relationship, of close or similar frequency, there will be a corresponding action/reaction. These reasoned or recorded reactions can be a manifestation that is known currently as hypnosis, telepathy, intuition, spells, mediumship, visions or precognition, as well as other less widespread forms, such as telekinesis, radiesthesia, clairvoyance, precognition and teleportation, as discussed in Chapter 8. As past humans could not understand how these phenomena happen, they always attributed them to imaginary beings, gods, angels, ghosts, geniuses, unconventional abilities and other similar factors. This human peculiarity of color sensitivities occurs in architecture, arts and decorations, and in all religions from the beginning of humanity to the present day [22–27].

REFERENCES

[1] R. Wilhelm, The Book of Changes, Illustrated ed. Dover, ISBN-10: 0486832589, ISBN-13: 978-0486832586, 14 Aug 2019, p. 480.

[2] D. Robson, The Intelligence Trap: Why Smart People Make Stupid Mistakes and How to Make Wiser Decisions. Hodder & Stoughton, Center for Practical Wisdom at the University of Chicago, opened in Jun 2016, p. 263.

[3] BBC News Brasil, www.bbc.com/portuguese/geral-57541472, accessed in 17 Jan 2025.

[4] D. Morehouse, Psychic Warrior: The True Story of the CIA's Paranormal Espionage Programme, 1st ed., Edited by Michael Joseph, ISBN-10: 0718141784, ISBN-13: 978-0718141783, 01 Jan 1996, 272 pp.

[5] P.E. Tetlock and Dan Gardner, Superforecasting: The Art and Science of Prediction. Library of Congress Cataloging-in-Publication Data, 1954.

[6] P.E. Tetlock, Expert_Political_Judgment. Princeton University Press, ISBN-13: 978-0-691-12871-9, 2005.

[7] BBC News, Como a ciência explica pessoas que parecem prever o futuro (BBC World News, How science explains people who seem to predict the future), https://g1.globo.com/ciencia-e-saude/noticia/2021/06/19/como-a-ciencia-explica-pessoas-que-parecem-prever-o-futuro.ghtml, 19 Jun 2021.

[8] D. Hawkins. Wikipedia, 1961. Manhattan District History, Project Y, the Los Alamos Project, Vol I: Inception until August 1945. Tomash Publishers. https://en.wikipedia.org/wiki/David_Hawkins_(philosopher), ISBN 978-0-938228-08-0. LAMS-2532. Retrieved 20 Feb 2014.

[9] N. Wiener, Generalized harmonic analysis, Acta Mathematica, 55(1), 117–258. https:// doi.org/10.1007/BF02546511, 1930.

[10] D. Robson, The Expectation Effect: How your Mindset can Change your World, Published in the US on. Amazon-Barnes & Noble-Bookshop.org-Apple Books, published in The United Kingdom by Canongate and in the USA by Henry Holt, 15 Feb 2022.

[11] J. Murphy, The Power of Your Subconscious Mind Paperback. Martino Publishing, Reprint of 1963 edition, ISBN: 978-1-61427-019-5, 31 May 2011.

[12] J. Murphy, Maximize Your Potential Through the Power of Your Subconscious Mind for an Enriched Life: Book 6, Google Books. Retrieved 04 Nov 2013.

[13] J. Olesen, 22 Aura Colors and Their Meanings: Learn How to Read Auras, 2013.

[14] Cristais Aquarius, Loja virtual de pedras e cristais para esoterismo terapias holísticas, decoração e coleções, www.linkedin.com/company/cristais-aquarius/about/, tel. 05515 32633894.

[15] Astrocentro, Maior plataforma de serviços esotéricos do Brasil, Portal Vida Livre, 05511 3164–0313, https://portalvidalivre.com/articles/3033, 2023.

[16] B.S. Farias and P.C. Landim, Inclusive graphic design for elderly (Design Gráfico Inclusivo para Terceira Idade), HFD, 8(15), 35–48, ISSN: 2316-7963, https://doi.org/10.5965/2316796308152019035, Mar 2019.

[17] M. Bernard, C. Lias, and M. Mills, The effects of font type and size on the legibility and reading time of online text by older adults. In: Conference on Human Factors in Computing Systems, New York, 2001, pp. 175–176.

[18] J. Olesen, 22 Aura Colors and Their Meanings: Learn How to Read Auras, www.color-meanings.com/aura-colors-and-meanings-how-to-read-auras/, 2013.

[19] J. Herding, Crystals: A Complete Guide to Crystals and Color Healing. Ivy Press, ASIN: b083qrw6fc, 01 Oct 2019, 320 pp.

[20] B. de Lara, A cura dos chakras com cristais: Manual Prático para Conquistar o Equilíbrio Emocional e Físico, Portuguese edition, Editor Pensamento, 1st ed., ISBN-10: 8531519853, ISBN-13: 978-8531519857, 25 Sep 2017, 96 pp.

[21] P. Mercier, The Chakra Bible, Patricia Mercier, Part of the Godsfield Bibles, UK ed. Godsfield Press, ASIN: 1841813729, ISBN-10: 9781841813721, ISBN-13: 978-1841813721, 02 Nov 2009, 400 pp.

[22] H. Motoyama, Theories of the Chakras: Bridge to Higher Consciousness. Theosophical Publisher House, ISBN-10: 0835605515, ISBN-13: 978-0835605519, 01 Nov 1981, 293 pp.

[23] C.W. Leadbeater, The Chakras: A Monograph, German ed., 8th ed. Theosophical Publishing House, ASIN: B0000BKRNZ, 01 Jan 1968.

[24] H. Johari, Chakras: Energy Centers of Transformation. Destiny Books, ISBN: 978-089281760-3, 2000, 168 pp.

[25] J.P. Miller, Le livre des chakras, de l'énergie et des corps subtils, Les secrets de l'énergie, Les Editions Quebecor, Les Éditions Québec-Livres, French edition, 30 Jul 2012, 160 pp.

[26] S. Karagulla and D.G. Kunz, The Chakras and the Human Energy Fields. Quest Book, 01 Apr 1989.

[27] B.E. Snow and H.B. Froehlich, The Theory and Practice of the Color. State College of Agriculture, Cornell University Library, 1920.

10 Historical Facts about the Human Mind

10.1 INTRODUCTION

This book addresses information processing in the sensitive brain, which begins with the stimuli generated by their bodily sensory organs such as the nose, eyes, smell, taste and the entire area of the skin. These stimuli, coming from the most diverse sources in the environment where human beings live, record information in the brain such as pain, pressure, smell, taste, colors, fears, among others. Each of these types of information reach the brain through electrochemical communication between neurons and then the brain gives a response after a certain amount of processing. To do this, neurons emit chemical and electrical signals to carry out cellular communications and, therefore, can easily be affected by the highly electromagnetic environment in which human beings live. Neuron signals are mostly stored chemically and communicated electrically.

The examples in this chapter clearly show the action of the human brain's receptivity to the most diverse types of information coming from the environment and the beliefs with which its bearer lives. This receptivity is the sum of information, beliefs, imaginations and sensitivities accumulated throughout a sensitive life that enable more or less accurate predictions of what is happening in a certain sector of human existence or in the lives of susceptible individuals. Such information can reach the brain of the so-called 'seer' through temperature, color, body movements, tone of voice, brightness in the eyes, similar scenarios, electromagnetic ambient, intonation of voices and emissions of natural frequencies from the bearer's body. It does not matter whether it comes from the brain or the body, previous experiences, the sensitivities of the receiver, ignorance of all that makes the mind more receptive/emissive and many other causes. It seems that all these manifestations have to do with what is believed out of fear of what could be and might happen.

10.2 RECENTLY OCCURRED FACTS

Chemical signals are involved in short-distance communication between nearby cells, while pulses of electrical current emitted by neurons are in long-distance communication. The long bodily distance for the chemical signal can be considered as the route that the signal needs to take from the brain to the muscles, signaling for the person

DOI: 10.1201/9781003604037-10

to produce actions such as blinking their eyes, scratching their head, taking a step or moving a toe.

When talking about frequencies beyond the usual ones easily perceived by the mind, others require explanations that are more complex. This is the case, for example, of visualizing colors that go beyond infrared or ultraviolet, as discussed in Chapter 2. This causes human beings to live in a surreal world where everything is based on impressions from the senses or all people always perceive hidden intentions, which are not perceived in the same way. With all of this, it becomes very difficult for a person to know what is true or false in ideas coming from histories, stories, religions, beliefs, ignorance, misinformation..... .

In this chapter, some examples are given of globally famous people who are sensitive and able to perceive other phenomena than those usually seen by the senses of ordinary people due to their sensory systems differing from each other or because they go a little bit further than mortals usually can. All of this seems to explain the reasons why certain people may go far beyond the usual limits.

10.2.1 BROTHER VITRÍCIO AND THE SENSITIVITY TO SUBCONSCIOUS INFORMATION

Luiz Benjamim Henrique Rech (Brother Vitrício) was a Brazilian religious member of the Order of Marist Brothers. Luiz was the son of Anna and Nicolau Rech and was born in Santa Cruz do Sul (RS – Brazil) in 1917, a descendant of German parents, and died in 2005 in São Lourenço (MG – Brazil). According to information extracted from the newspaper *A Razão* (11 November 2005) from Santa Maria-RS, he created the technique called 'lethargy', having lived his last years in Minas Gerais [1].

Between the 1950s and 1960s, Brother Vitrício was in the news in Brazilian and foreign newspapers, magazines and books, having been invited by hundreds of health professionals around the world to give courses and lectures on the subject. He was the author of a controversial book since its release (1956), entitled *Don't Read....*, the first work to address lethargy. He wrote of "My intention to explain the mediumistic phenomenon such as lethargy and scientifically demoralize spiritualist sessions, which claim to be consulting souls from the other world. Spiritualist mediums do nothing supernatural, but only capture, through lethargy, the thoughts of close people, very much alive. They generally capture the thoughts of the living son who wants to speak to his deceased mother. The son thinks a lot about his mother and the medium captures messages, not from the mother to the son, but vice versa".

Brother Vitrício and Paulo Paixão were the biggest promoters of hypnosis in Brazil, especially the lethargy technique. Paulo Paixão is the author of several books talking about this subject that have been translated into different languages and is a figure present in a good part of the serious literature on hypnosis around the world.

Brother Vitrício gave many lectures and made practical demonstrations of lethargy in the main hall of the Colégio Arquidiocesano in São Paulo. He always stated that his purpose in presenting himself to the public was to prove that spiritualist phenomena and 'miracles' occurring in macumbas have perfectly natural causes. He recalled experiments carried out in the USA and other countries under scientific control, such as those carried out by scholars of extrasensory phenomena talking about the generalities of acupuncture. This ancient Chinese technique for treating illnesses uses the fixation of needles in certain regions of the body, whether gold, silver or platinum needles. It thereby states that the capacity for extrasensory perception is latent in all human beings and that its phenomena can be triggered, according to certain techniques that can be physical and/or psychic, ranging from moderate trances to clairvoyance. Brother Vitrício said that the vaunted intervention of spirits is fully explainable by the new science [2,3].

Among his abilities, Brother Vitrício said that he did not need to use hypnotism or suggestion to promote lethargic states in people who underwent his experiments. People who underwent his experiments immediately fell into a lethargic state caused by simple touches on the back, vertebrae or forearm. Soon after, the patient appeared completely insensitive to physical stimuli, including contact with the cornea by the operator's hand, which inserted needles into the cheeks without causing them to bleed, and some other similar experiments. He also demonstrated that patients were as if taken by spiritual incorporations, even though they were simply in lethargy. Using music and playing in the labyrinth and in the semicircular canals of the ears, he could make several young people of various ages fall asleep who, in addition to being asleep, were made insensitive to pain.

10.2.2 MUSICAL HALLUCINATIONS

Tony Cicoria was 42 years old, strong and in great shape. He played football in the college and became an orthopedic surgeon in a small town in upstate New York. On a pleasant autumn afternoon, he went to a family reunion, in a pavilion on the edge of a lake. There was a breeze, but he noticed some storm clouds in the distance. Rain would come, he thought. Before the age of cell phones, in 1994, he went to the public telephone near the pavilion to make a quick call to his mother. Tony still remembers every second of what happened to him next: "I was talking to my mother on the phone. It was drizzling and thunder could be heard in the distance. My mother hung up. The phone was about a foot away from where I was when a flash of light coming from the device hit me, I remember. It hit me in the face. My next memory is of flying backwards" [4,5].

The lightning that struck Tony Cicoria had a curious effect on him after the accident, with him developing an insatiable desire to listen to music played on the piano. Strangely, this had never aroused his interest before, but it ended up making him a well-known pianist and composing his own melodies. One of them was the melody *The Lightning Sonata*. With this ability, Cicoria ended up becoming the character in *Hallucinations*, written by neurologist Oliver Wolf Sacks (2013). This Anglo-American writer and amateur chemist was a renowned professor of neurology and psychiatry at Columbia University, where he earned the honorary title of 'Columbia Artist'. In his works, he brought together several clinical cases that illustrate the relationship between the brain and music.

The accidental case of lightning and similar can offer a neuroanatomical explanation for the sudden revelation of an orthopedist's musical talent like that of Tony Cicoria. "Analyzing him from a neurological perspective, I think that his brain must now be very different from how it was before being struck by lightning or compared to what it was like in the days immediately following the incident, when neurological exams had not detected any major problems … Changes were presumably occurring in the subsequent weeks when his brain was reorganizing itself preparing, say, for musicophilia" [4,5].

From his point of view, Tony Cicoria considers himself a very different person after his lightning accident. He reveals that after putting the dream song on paper, he felt a certain feeling of freedom, even going so far as to resume his marriage: "Now there is balance, where there was no balance before. Do I still play all the time? I touch, but I try to make it not invasive. We spend time together every night when I get home, and if I don't play that day, it's not the end of the world like it used to be".

10.2.3   THE POLYGLOT MAN

Edvalson Bispo dos Santos from the city of São Francisco, north of Minas Gerais, Brazil, 60 years old, nicknamed "dizzy chicken", was very poor and had little education, but he began to speak English, Japanese and German in addition to Portuguese. As he says, he learned these languages in his dreams, while sleeping, as he only studied at a public elementary school, only having poor knowledge of how to read and write. As a boy, Edvalson begged for food in homes. In one of them, he received food from the house cleaner and when the boss saw this, she kicked the boy's plate. That night, the boy was deeply hurt, and went to bed early. He used to listen and tell stories sitting on the sidewalk near his house and after the humiliation he went through, it all ended for him [5].

Edvalson learned all about some foreign languages at the age of 7 after being deeply humiliated by that lady in his hometown. One night he dreamed that he was sitting on a log on the beach by the sea, very sad, with his head down, wondering why that woman had done that to him. Suddenly three boys dressed in white appeared, which he considered to be angels, as the boys motioned for him to raise his head. One of them approached him and spoke in Japanese, saying he was Toshio. The Japanese began to write the phrase: "God is love". Then a second boy introduced himself with his name saying he was a German named Hans and saying in German "nice to meet you", also saying, "God is love". This boy walked away and then a third boy came saying he was English called Paul, also saying, "God is love". Edvalson woke up impressed and terrified right after the dream, wondering if the dream was real or not. He looked for his mother and asked, "Mom, where are the three boys who were on the beach"? To which she replied, "You weren't on the beach, you were here in your room!" The boy then became all confused. He was insistent to his mother, but she denied everything. He did not accept it and insisted on wanting to know where the three boys were, looking for them throughout the house.

Edvalson's dream repeated itself for three days. The mother then decided to take the boy to the priest at the rectory of the Catholic church. She told the priest: "I brought this boy here because he has been saying some things as that he saw three boys and he is no longer talking to me; I would like you to give him a blessing because everything seems a bit convoluted". The priest, who was German, then came face to face with the boy and asked him in German, "Wie geht es dir?", "How are you?" in German. Then the priest asked the mother, "How do you explain this to me?" She responded: "I think the boy is going crazy". The priest took the boy away and asked the mother, "Do you believe in the Holy Spirit?" To which she replied, "Everything that comes from God I accept!" The priest then says: "That's what's happening to your son" [6,7].

The dreams continued to happen for the boy for about 15 years. The finger used by the boys in their dreams was as if it were the chalk to write on and the sand on the beach was the blackboard. The boy decided to memorize everything and only learned Portuguese well when he was 11 years old. Before knowing Portuguese well, he said he already knew foreign languages. He learned in his dreams both the pronunciation and grammar of the three foreign languages. The three dream boys taught Edvalson, who decided to buy a notebook to write down the lessons and explain to his mother the meanings of the words in the languages he was learning.

The boy was poor and suffered hunger as a child. For some reason, the boy, now grown up, one day went hungry, childhood dreams were repeated after so many years, and he, now 60 years old, saw the boys again at the same age he met them when he was 7 years old.

Edvalson, now grown up, went to the Christian Congregation in Brazil (CCB) and today he supports his family. His story was presented on the Brazilian television network Record. Later, he was tested by three teachers in Japanese, German and English and apparently did very well. Nowadays, he teaches needy children and continues to live a poor life. In addition to TV Record, he has given interviews on several other Brazilian television channels.

10.2.4 SCIENCE AND SUPERNATURAL PHENOMENA

Few people have not had the frightening feeling that there was a presence in their room, even though they were sure they were alone, and perhaps they are reluctant to admit this experience. Perhaps, it was something profound that they wanted to share with other people, or even more likely, an experience may have fallen somewhere between reality and these extremes. Unless there was an explanation to help process experiences like these, most people have difficulty understanding what happened. Current studies are demonstrating that this ethereal experience is something that can be explained using scientific models of the mind and body and the relationships between them.

One of the largest studies on strange presences was carried out in 1894. A British entity called the Society for Psychical Research (SPR) published its *Census of Hallucinations* that year [8]. This document describes research that involved more than 17,000 people from the UK, the USA and mainland Europe. The objective was to find out how common it was for people to receive seemingly impossible 'visits' that foreshadowed death. The SPR concluded that these experiences were too common to be the result of chance, since out of every 43 people surveyed, 1 had had such an experience. Among the society's patrons were former British Prime Minister William Gladstone and the poet Alfred, Lord Tennyson.

In 1886, the SPR published *Phantasms of the Living*, documenting 701 cases of telepathy, premonitions and other unusual phenomena. An example reported in this work was the case of the Reverend P.H. Newman, from Devonport, in Plymouth

(England), who reported a visit to New Zealand, where a nocturnal presence advised him to cancel his ship trip that was scheduled for the morning of the following day. The reverend later learned that all the passengers on that trip drowned. At the time, ghost stories were criticized for being unscientific. After the SPR study, the census was received with less skepticism, even though many responses were avoided, as only those who had something to say would bother to respond to the survey. Of note, experiences may occur in homes all over the world and contemporary science offers some ideas for understanding them.

In the Victorian era, the presences documented by the SPR were often benign or comforting. In modern examples, such presences caused by sleep paralysis usually emanate evil. Examples range from the Fradinho da Mão Furada, in Portugal, who managed to infiltrate people's dreams, to the Ogun Oru, from the Yoruba ethnic group, in Nigeria, who believed that his victims had been bewitched.

Some sleep paralysis researchers have focused on specific characteristics of people who wake up in strange situations after sleep. Most of these people find sleep paralysis a frightening experience, even without the hallucinations. In 2007, sleep researchers J. Allen Cheyne and Todd Girard argued that if a person woke up vulnerable and paralyzed, their instincts would make them feel threatened and their minds would fill in the blanks with images: if the person felt like prey, this was because there must be a predator. Another approach is to observe common characteristics between visits during sleep paralysis and other types of presence.

Research has shown over the last 30 years that the presences are not only frequent in the hypnagogic setting because they have also been reported in cases of Parkinson's disease, psychosis, near-death experiences and bereavement. These findings indicate that it is unlikely that this is a sleep-specific phenomenon.

It is known through case studies in neurology and experiments with brain stimuli that presences in sleep can be caused by indications from the body. In 2006, for example, a woman whose brain was electrically stimulated at the left temporoparietal junction (TPJ) perceived neurologist Shahar Arzy and his colleagues managed to create a figure of that. A figure appeared to mirror the position of the woman's body and the TPJ combined information about the human senses and their bodies.

A series of experiments also demonstrated in 2014 that undoing people's sensory expectations also appears to induce a sense of presence in healthy people. In these experiments, researchers 'trick' participants into feeling as if they are touching their own back by synchronizing their movements with those of a robot directly behind them. The human brain perceives the synchronization, deducing that the person is producing that sensation, and when the synchronization is interrupted (causing the robot's touch to be slightly uncoordinated), people can suddenly feel that another person is present: a ghost in the machine [10].

Changing the sensory expectations of a situation induces something similar to a hallucination. This logic can also be applied to situations such as sleep paralysis. All the usual information that a person has about their body and their senses is destabilized so that the feeling that there is 'another person' nearby is not surprising. The person may feel as if they are another presence, but in fact, they are themselves in their imagination [11].

In a 2022 survey, an attempt was made to trace the similarities between the presences observed in clinical cases, spiritual practices and endurance sports, activities known for producing a series of hallucinatory phenomena, including presences. In all of these situations, several aspects of the sense of presence were quite similar. The patient felt, for example, that the presence was directly behind. The three groups described presences related to sleep, but also presences caused by emotional factors, such as grief and loss.

Despite having emerged centuries ago, the science of felt presence is actually just beginning. Such scientific research may provide a more comprehensive explanation or require several theories to clarify all these cases of presence. However, it must be said that the encounters described in the book *Ghosts of the Living* are not echoes of a bygone era. If someone has not had this disturbing experience yet, he/she probably knows someone who has.

10.2.5   LETTERS FROM CHICO XAVIER

Psychography is the technique used by mediums to write texts under the influence of what they consider a 'disembodied spirit'. To do this, they use their own hand, known as 'direct psychography' or 'manual psychography'. Of all the forms of 'spiritual' communication, handwriting is the simplest and most complete because it allows anyone to establish permanent and regular relationships with what are considered 'spirits' [12].

A researcher from the State University of Londrina, Paraná, Brazil, Carlos Augusto Perandréa, studied 400 letters psychographed by Chico Xavier during his mediumistic trances, using the same techniques used to evaluate signatures for banks, police and the judiciary, known such as graphoscopy. This researcher compared the psychographed letters with the standard handwriting letters of the individuals before and after their deaths, concluding that they all had graphic authenticity.

In Brazil, psychographic letters occupy a prominent place among literate practices in spiritualist doctrine. Parents and family members who, upon receiving letters attributed to their loved ones, began to dedicate themselves to charitable activities as a way of overcoming the pain of their grief and consolidating their beliefs founded many of the well-known spiritualist institutions.

In the first decades of the 20th century, the leaders of the parapsychological research organization, the SPR, collected and considered authentic several messages psychographed by various mediums who attributed authorship to the spirit of F.W.H. Myers, an intellectual who was one of the founders of the organization. The managers also found that the psychographed messages had a very strong correlation of continuity between them, forming a kind of 'puzzle'.

More recently, in 2008, scientists from Thomas Jefferson University, together with the University of São Paulo, the Federal University of Juiz de Fora, the Federal University of Goiás and the University of Pennsylvania, using modern neuroscience resources, carried out scientific research in the USA to measure the brain activities of ten healthy Brazilian mediums while they were psychographing. Scientists found that during psychographic trances, those areas of the mediums' brains that the usually least activated areas of the mediums' brain were the ones that became most activated. On the other hand, anyone writing normally in a trance state, the areas related to reasoning, planning and creativity remain unchanged. It was noted that the psychographed texts were more complex than those produced in a normal trance state were. As the research recorded in the psychographed texts, the mediums produced mirrored messages, that is, they wrote backwards, wrote in languages they did not speak well and correctly described the ancestors of the scientists that they themselves said they did not know, among other things. For such scientists, the research results are compatible with the hypothesis that the psychographed text is not authored by the people physically writing, but rather the 'communicating spirits', as the mediums state. One of the other common points observed in such mediums was that they are huge admirers of Chico Xavier.

In 1990, the Medical-Spiritualist Association of São Paulo, Brazil, carried out research on 45 letters psychographed by Chico Xavier and considered authentic by the recipients, concluding, "The evidence of the spirit survival is very strong. Life is a fatality, according to the testimony of these 45 companions who exposed themselves completely, revealing the nuances of their personalities through the humble hands of the mediator".

Psychography has played an important role in the courts. The most famous case was undoubtedly that of Humberto de Campos, to whom in 1937, three years after his death, several chronicles and novels were attributed and began to be psychographed by the Brazilian medium Chico Xavier, Pedro Leopoldo city, state of Minas Gerais, Brazil. Among the works, all published by the Brazilian Spiritualist Federation, the most famous among Brazilian spiritualists was *Brazil, Heart of the World and Homeland of the Gospel*.

In 1944, Humberto de Campos' widow went to court, filing a lawsuit against the Brazilian Spiritualist Federation and Francisco Cândido Xavier, in order to obtain a declaration, by ruling that this mediumistic work "was or was not from the 'Spirit' of Humberto de Campos". If so, may she obtain the work's copyright. The subject caused a lot of controversy and, for a long time, occupied space in the country's main periodicals. The author of the case, D. Catarina Vergolino de Campos, was judged in need of the action proposed by a sentence of 23 August 1944, by Dr João Frederico Mourão Russell, acting judge at the 8th Civil Court of the former Federal District. She appealed this sentence, but the Court of Appeal of the former Federal District upheld it on legal grounds, with Minister Álvaro Moutinho Ribeiro da Costa serving as rapporteur.

In Brazil, in some cases, psychography has been used as evidence in court. Texts written by Chico Xavier were accepted as legal evidence in addition to others that were also presented by the defense. This evidence proved to be decisive elements in the sanctions applied in three cases of international homicide trials, which took place in the Brazilian states of Goiás, Mato Grosso do Sul and Paraná between 1976 and 1982 [12].

One of the most recent cases of psychography used in courts was registered in May 2006, in Porto Alegre (RS), with the defendant, Iara Marques Barcelos, being cleared of the murder of her ex-lover, Ercy da Silva Cardoso, thanks to a letter which would have been dictated by the deceased. More recently, on 17 May 2007, the trial of the defendant, Milton dos Santos, for the murder of Paulo Roberto Pires, known as

"Paulinho do Estacionamento", in April 1997, was suspended due to a letter received by the medium Rogério Leite in a spiritualist session held in 2004, in which Paulinho acquits the accused. However, lawyer Roberto Selva da Silva Maia indicated in an article that psychographed documents could be accepted in court as a private document, but not as judicial evidence. According to him, this is because Brazilian law establishes that death extinguishes human personality. Therefore, a dead person could not generate a legal document. Also according to him, psychography depends on the acceptance of religious premises, and the judiciary is not religious since the Brazilian state is secular and there would be no way to take advantage of the contradictory principle and broad defense [11].

Roberto Selva da Silva Maia had his opinion contradicted by lawyer Michele Ribeiro de Melo, who wrote her master's dissertation defending the acceptance of psychography as judicial evidence, having earned a master's degree in law from the Centro Universitário Eurípedes de Marília (Univem), Marília, São Paulo. According to her, "Psychography can be used as judicial evidence without violating any constitutional precept or procedural principle. Quite the contrary, the admissibility of this type of evidence occurs in compliance with the fundamental guarantee of the right to evidence, constitutional principles and the principles that govern the evidence in our legal system" [12]. Michele Ribeiro verified that the psychographed evidence does not offend the principle of the secular state, which provides for freedom of religious beliefs and cults. She assured that psychography, as a mediumistic phenomenon, is a natural faculty of the medium human being, studied by science and is not a religious element.

10.2.6 RUSSIAN BOY ON MARS

The 11-year-old Russian boy Boriska Kipryianovich living in Vulgograd claimed that he had lived on Mars even though the planet's atmosphere is made up mainly of carbon dioxide. Furthermore, Mars' gravity is 38% of what people are used to on Earth and that planet has been constantly bombarded by radiation. The boy from Mars said that he had arrived on Earth shortly after dying on his old planet [13].

In 2007, Boriska told impressive facts to scholars and journalists who, to this day, have not been able to demystify them plausibly. The boy assured people that he had lived on Mars in a past life, giving surprising details about his experiences on the red planet. His mother on Earth was a doctor and said he was a genius, as he began speaking just a few months after birth. At the age of one and a half, he was already reading, drawing and painting. In kindergarten, when he was 2 years old, teachers noticed his extreme ability to write and learn languages and that he talked a lot about Mars.

When he was just 3 years old, Boriska started talking about things he did not understand very well, related to the positioning of the planets and the mysteries of the Universe. As his classmates played in kindergarten, he talked about Mars as someone who had deep experience of living on the planet. His parents, however, stated that they never taught him anything about this and that he had an IQ of 200. Among his stories, he said that he visited Earth frequently when he lived on the red planet aboard ships commanded by the Martian Air Force. He was hardly challenged because he talked about details of the planet Mars that no one could refute. He said that Martians are extremely tall, over 2 meters, and that they stop aging at 35 years old. He also said that he was not the only one on Earth and that there were many reincarnations on Earth of the Martian race called indigo children. He also said that Martians could travel through time and that their race had been decimated thousands of years ago by nuclear wars. Such wars broke Mars into two sets of Martians and some survivors are still living on Mars. He said that they have learned to produce new weapons and that these survivors can be found during interspatial travel.

Boriska reported that he and other Martian boys came to Earth to warn Earthlings about the dangers of nuclear wars. He said he was reborn on Earth with the special mission of issuing a general warning. Based on this, he made frightening predictions that the Earth's magnetic centers would melt and, with this, the poles would change position with immediate effects on the environment with drastic meteorological changes. Similarly, he says, such factors occurred on Mars 42 thousand years ago, leading the planet to extreme climatic situations. With this, the lives of mammals and some of humanity's closest relatives were extinguished. The explanation for this is that as the magnetic field weakens, more cosmic rays penetrate the atmosphere and more solar radiation bathes the Earth. Extreme weather conditions will be caused on Earth by the reversal of the poles resulting in high temperatures and many storms.

Boriska talked about the sphinxes of Egypt saying that they contain an opening in the back and a secret. He also said that if they were opened, life on Earth would never return to normal, so no one should go near them. The key is hidden behind one of the ears. He also said that more children from there would come to Earth to help humanity out of its impending doom. Some people speculate that indigo children will be the next step in human evolution. The boy predicts global floods and that few humans would survive them.

As of January 2025, Boriska is around 27 years old. It is strange that no one knows where Boris is these days and that he is probably in the custody of President Vladimir Putin. Furthermore, his parents have also become recluses, and the neighbors have heard little from them since they eventually moved away [13].

10.2.7 BOY WONDER ON THE PIANO

Pianist Gavin George was born on 10 May 2003, in Westerville, Ohio, USA, began formal instruction at the age of three and a half and made his concert debut at the age of seven, performing Haydn's Concerto in D major. A gold medal winner two years in a row at the American Association for the Development of the Gifted and Talented International Piano Competition, George has performed numerous times at Carnegie Hall and received a full scholarship to study and perform at the Vianden Music Festival in Luxembourg

Gavin George became the youngest musician when he performed on the Emmy-winning show *From the Top*, broadcast on National Public Radio, Inc. (NPR). He also received first prize in the SAA International Piano Concerto Competition performing Beethoven's Concerto No. 3. Sponsored by PNC Bank, *Gavin George: Live on Your Screen* broadcast his performance of Rachmaninoff's Concerto No. 1 with the Westerville Symphony Orchestra. It was chosen by the Cleveland Women's Orchestra to be performed in their 79th Anniversary Concert, where they performed Mendelssohn's Concerto No. 1 at Severance Hall, Cleveland. He was selected as one of the documentary profiles in the PRODIGIES series appearing on the YouTube channel THNKRTV. He was also an actor on both the CBS *Early Show* and the *Queen Latifah Show*, and featured in the online *National Geographic* magazine article, "Exploring Characteristics of Prodigies". More recently, Gavin performed excerpts from Rachmaninoff's Rhapsody on a Theme by Paganini with the Detroit Symphony Orchestra [13,14].

George gave national and international recitals and concerts from a young age. Among his orchestral performances are:

- Mendelssohn Piano Concerto No. 1 (Ashdod Symphony Orchestra, Columbus Symphony Orchestra, Dayton Philharmonic, Lakeside Symphony Orchestra);
- Schumann Piano Concerto in A Minor (Springfield Symphony Orchestra);
- Ravel Concerto in G Major (Lexington Symphony Orchestra);
- Mozart Piano Concerto No. 21 (ProMusica Chamber Orchestra & Dayton Philharmonic);
- Beethoven Piano Concerto No. 3 (Newark-Granville Symphony Orchestra);
- Rachmaninoff Piano Concerto No. 2 (Firelands Symphony Orchestra);
- Mozart Piano Concerto No. 5 (Mainly Mozart Festival Orchestra).

His recital venues include:

- Constitution Hall (Washington, DC);
- Ocean Reef Club Performance Center (Key Largo, FL);
- Wexner Center for the Arts (Columbus, OH);
- Mainly Mozart Festival (San Diego, CA);
- Shalin Liu Performing Arts Center (Rockport, MA);
- Xavier University (Cincinnati, Ohio);
- Orchestra Hall (Minneapolis, MN);
- Tassel Performing Arts Center (Nebraska);
- Gartner Auditorium (Cleveland, OH);
- Vianden Castle (Luxembourg);
- Sorrento (Italy).

George studies piano with Dr Sun Min Kim and was a student of Antonio Pompa-Baldi. He participated in masterclasses with Menahem Pressler, Ann Schein, John Perry and Boris Slutsky, among others. In addition to his musical endeavors, Gavin is a National Merit Scholar [14].

10.2.8 MIRACLES OF MOTHER TERESA OF CALCUTTA

Information from the Archdiocese of Rio de Janeiro, the Holy See Press Office, newspapers *Zero Hora* and *Zenit* report that the first miracle attributed to Mother Teresa, founder of the Missionaries of Charity, was the healing of a 30-year-old woman who suffered from an abdominal tumor in Bangladesh. This miracle allowed her beatification in 2003 [15–17].

Pope Francis canonized Mother Teresa of Calcutta, who was born Agnes Gonxha Bojaxhiu in Uskup, Ottoman Empire (now Skopje, North Macedonia), on 26 August 1910. She was declared in a ceremony to be a saint, motivated by her intercession and the recognition of a miracle attributed to her. The miracle happened to São Paulo engineer Marcílio Haddad Andrino, then 34 years old, who now lives in Rio de Janeiro. This engineer was cured of eight brain abscesses in 2008. Earlier that year, Andrino had some strange symptoms, such as double vision and lack of balance, to which he gave little thought. After a few months, he had a more severe seizure and fainted. After that went to several doctors who were unable to find out what his problem was. Among the symptoms he experienced was the loss of most movements on the left side of his body and he had difficulties in the cognitive area. His fiancée's boss, Fernanda Rocha, suggested that the couple pray to Mother Teresa, assuring that he himself had been cured of an aneurysm through her intercession. They began to pray, but at first the situation only got worse. Their wedding was scheduled for September 27. On the fifth, the day of the liturgical celebration of Blessed Teresa of Calcutta, Fernanda spoke with her parish priest, in São Vicente. This parish priest that day had celebrated mass at the Missionaries of Charity house, in Santos-SP, and received a relic from the nun who gave it to Fernanda, recommending that the couple once again ask for Mother Teresa's intercession [14,15].

Fernanda and Andrino got married on the scheduled date even though he was very weak and had great difficulty getting around. The following month, he had an even more severe seizure and was taken to a hospital in Santos. Only then did a doctor warn them that the situation was extremely serious, as there were three major and five secondary abscesses in the engineer's brain, finally diagnosing Andrino. He started taking antibiotics aimed at reducing the abscesses.

Every day, the couple Fernanda and Andrino prayed to Mother Teresa in front of the relic presented by the parish priest. Ten days after hospitalization, however, the left side of Andrino's body became completely paralyzed. In early December, he woke up with an unusual headache. Several doctors rushed to his room and Andrino was immediately sedated, but not before asking his wife to continue praying and asking his family and parishioners to also pray for him. Fernanda, at home, prayed fervently to the 'saint of the gutters'. When the engineer woke up, he was in the operating room asking the doctor "What am I doing here?" saying that he felt very good. The doctor then decided to postpone the surgical intervention until the next day and the patient was able to sleep well, without pain. The next day, doctors performed tests while he was sedated. They discovered that there was a drastic 70% reduction in the abscesses present in his brain. The surgery was cancelled. Three days later, another exam was carried out, certifying that there was no longer any abscess. A few days after the tests, Andrino left the hospital, now able to walk, although with some difficulty, and spent Christmas with his family. Six months later, he returned to work, without any sequelae.

Newly married, Andrino and Fernanda wanted to have children, but doctors warned him that the chance of having children was less than 1% due to the strong medications taken during treatment as well as others a few years earlier, due to a kidney transplant. A month after returning to work, however, they discovered that Fernanda was pregnant and, three years later, in 2012, the couple had a second child. One of the doctors who analyzed the miracle, Carlo Jovine, said at the time: "There is no precedent for a single brain abscess to be cured, but with eight brain abscesses and acute hydrocephalus, the death rate is practically 100%. From this chain of events and clinical examinations, specialists and experts necessarily concluded that they were dealing with a scientifically inexplicable event, which occurred in a decisive, immediate, lasting and total way. This, for the Church, is equivalent to a miracle" [15,16].

Andrino said in an interview with the Archdiocese of Rio de Janeiro, "After the miracle, my faith increased a lot; I believe that there is a God who watches over each of us. I see God in all things, even the small ones. Before, I could not take a step, today I am grateful to be able to walk. I see that everything has the hand of God, even in simple walking".

10.2.9 POST-DEATH EXPERIENCES

Recently, an unprecedented scientific study showed that a person's consciousness does not die immediately when their heart stops beating, but that they spend their lives before their eyes or have the sensation of leaving their own body and these are not hallucinations. Some of the elements that make up the popular imagination about what happens when someone dies refer to darkness, the end of pain, the exit to the light and then a feeling of peace [18–21]. Religions have their answers, writers explore the topic creatively, people have their beliefs, but what really happens?

A study conducted by the Grossmann School of Medicine, New York University, in the USA, concluded that one in five people who survive cardiopulmonary resuscitation after a cardiac arrest could describe experiences of lucidity about death that occurred while they were apparently 'dead' and without a heartbeat.

Throughout history, death has been observed based on the social convention that there was an insurmountable line between life and death and that, once anyone crossed it, there was no return, as explained by Sam Parnia, director of the study presented in recent 2022 Scientific Sessions of the North American Heart Association, in Chicago, USA. "In the last 60 years, this concept has been called into question because the discovery of cardiopulmonary resuscitation made it possible to restore life to some people who, from a biological point of view, had entered death", according to him. "These types of people have been reporting experiences for over 60 years and there are millions of them around the world who have recounted the same experiences". However, for years, these stories were considered simple hallucinations, tricks of the brain, or drug-like experiences. However, this current research has demonstrated that this comparison is erroneous.

Researchers studied 567 people who received cardiopulmonary resuscitation after a cardiac arrest during their hospitalization between May 2017 and March 2020, in the USA and UK. Of these, less than 10% survived. "It is necessary to understand that cardiac arrest is not a heart problem, but just a medical term to designate death", says Parnia [18].

Of the group analyzed, 85 people were studied with optimal brain monitoring, the largest group ever researched to date. Researchers also faced the challenge of installing all the necessary medical mechanisms to monitor the brain. To conduct this study, they used on the one hand cerebral oximetry, which is a non-invasive technique to monitor oxygen changes in cerebral metabolism based on near-infrared spectroscopy technology. In it, near-infrared photons are emitted into the patient's frontal skin. On the other hand, a portable electroencephalogram device was used. "Cardiac arrest is an emergency as it occurs very suddenly and without warning", says Parnia. "Typically, teams need to arrive within five minutes, enter the emergency and get all the devices working. So collecting the data is actually a challenge". Parnia is director of the Parnia Lab, the world's first research laboratory dedicated to improving resuscitation care and exploring what happens in the human mind during and after a cardiac arrest. Previous studies on animals had shown that they show waves of electrical activity in the brain at the exact moment after cardiac arrest.

Another study on brain activity, presented in February 2022, analyzed the brain activity of a woman at the exact moment of her death. A sudden increase in what is called gamma brain activity was observed, which are the waves activated when a conscious person retrieves memories and mentally processes this information. With this background, Sam Parnia's team aimed to answer two questions: 1) what are people's experiences when their heart stops beating and they are revived and 2) whether it is possible to find brain markers that confirm the reports of people who claim to have experienced lucid consciousness. Nevertheless, most of all, they sought to distance themselves from the term 'near-death experience'. In the scientist's opinion, this expression has been misused throughout history to describe countless types of occurrences that have nothing to do with death and are not even similar to each other. "Some people use the expression 'near-death experience' to talk about dreams. Others, to talk about drug use", he says. "For us, these are real death experiences. First, because the hearts stopped beating and, second, because people realize that they have died when they return".

Often, when people are resuscitated by cardiopulmonary resuscitation, they remain in a coma for days or weeks. This lapse of time could cause countless memories and the research tried to differentiate the types of memories that are formed. "These people can describe all sorts of different things that have been mistakenly called near-death experiences, but they are probably distinct", explains Parnia. Therefore, the researchers separated the two groups. "We concluded that there are clearly different experiences that occur in the days and weeks after resuscitation, usually when the person is beginning to awaken from the coma, so that they have nothing to do with the death experience", he says. Even so, the research ruled out that these were other experiences, such as dreams. "Everyone has random dreams, which are all different", according to Parnia. "But with the experience of death, people mention five main themes, even if they don't know each other, and these themes are wonderfully grouped together". These groups are evaluation of life, feeling of returning to the body, perception of separation from the body, perception of heading towards a destination and returning to a place that is perceived as home. This was the first part of the study. "In this way we were able to demonstrate that, essentially, the experience of death is not the same as hallucinations, delusions or dreams", explains the researcher [18,21].

The second step in Parnia's studies was to install brain monitors on people to look for these brain markers of lucid consciousness. This is how researchers discovered that, up to an hour after receiving cardiopulmonary resuscitation, there were signs of high-level brain activity, called alpha, beta, theta, delta and gamma waves, as explained in Chapter 5. "Some of these waves are consistent with what happens when we have conscious thought processes, when we are analyzing things, reliving life, memories, and when we have higher-order consciousness", explains Parnia. "In this way we were able to demonstrate, for the first time, brain markers of the lucid experience of death. In addition, of course, to their own experiences".

In fact, no one expects people to remember anything when they are resuscitated. However, that does not mean they did not have the experience. Due to sedative medications, deep coma and inflammation of the brain, which is the first occurrence seen when the heart starts pumping blood again, it is normal for people to forget everything. "We will never be able to get 100% of people to remember everything, which has a lot to do with the effect of the brain and the administered medications", explains the researcher; "39% of people have vague memories but can't remember the details, and 20% have what we call a kind of transcendent experience, 7% remember hearing things and 3% remember seeing something", he details.

Among the group of people who remembered what they experienced during the period in which the heart stopped pumping blood, but their brain continued to record markers of elevated brain activity, the study collected several experiences, of unknown duration. "It could just be a few seconds, I don't know", Parnia acknowledges. Among the statements collected in the study, several patients recalled having evaluated their lives and made statements such as:

- "I looked up and saw my destiny";
- "It wasn't that I was in a tunnel. It was as if a tunnel was created around me due to the incredible speed of my travel";
- "I went through a tunnel at great speed. It was wonderful and I didn't want to go back";
- "I knew I was home";
- "I wanted to go to the light. I wanted to come home";
- "I reviewed my life and, during this review, I saw scenes from my life again".

Others claimed to have experienced a separation from the body; others, the sensation of returning to the body. Others had the perception of heading to a destination and returning to a place that they felt was home:

- "My whole life flashed before me ... at first, it was very fast. Then, a few moments slowed down. Everything was shown to me, everyone I helped and everyone I hurt";
- "My life and all its events began to reproduce themselves in my mind, but in a very clear, real and vivid way".

For Parnia, the interesting thing is the different aspects of the life review. "Normally, we remember 1% of our entire lives when we are alive. But somehow it is remarkable that, in death, people remember everything even though their brain is turning off". "But, interestingly, it's not like a movie, as it erroneously appears in the media", he explains. "It's a very deep, intentional, meaningful reevaluation of everything they've done and said and thought. They judge themselves, they judge their actions based on their morals and ethics, which is really admirable". "And all of this happens when they're experiencing death, which, again, is very interesting. And that's what makes it impossible for this to be a hallucination". Parnia adds. "They know they're reliving everything spontaneously, which is fabulous".

Patients in the Parnia study said they felt "terribly bad", for example, when they experienced the pain they caused to others. However, they also felt the same joy and happiness that their actions brought to those close to them. At this point, Parnia explains that it is important to take into account that, normally, in order to carry out daily life, not all aspects of the brain be processed because it would be unbearable. "Your brain is active in certain parts that are important and others are usually inhibited with a kind of disruption system that serves as a brake", he explains. "What's interesting is that with death, what we're seeing is that as people go through death, the brain shuts down, it loses speed, and when that happens, the disruption systems are eliminated and the inhibition process is suspended", says Parnia about the process that they were able to prove with markers that measure the electrical activity of the brain. With this, it was possible to observe activity in parts of the brain that normally cannot be accessed. "Everything that happened in their lives is recorded and people are able to relive it, which is absolutely remarkable", he said. "It's definitely a real death experience and now we understand it because we're looking at it from a scientific point of view, but also from an evolutionary perspective". "Why is it that when you die, all the things that matter to you, like paying your bills, your mortgage, your dinner, your job, whatever ... completely disappear?" asks Parnia. "They are no longer important. What is evident, what matters, in fact, and what stands out in your mind at the time of death, is your conduct as a human being. The moral and ethical aspects of your actions and that is really fascinating", concludes the researcher [20,21].

10.2.10 VOTORANTIM'S PRODIGY CHILD

According to the family, who live in Votorantim (SP), these are some of the many skills developed spontaneously by the 2-year-old boy Samuel Brito: reading, doing calculations, memorizing the names of planets and even knowing some colors in English. The boy's mother, Camila Brito, says she began to notice that her son had unusual abilities when she saw him correctly fitting together the pieces of a toy with different geometric shapes. Afterwards, she and her husband, Felipe Brito, discovered that the boy knew the colors and, once again, were surprised when they handed him an alphabet book and saw that he also identified the vowels [22]. "At a year and a month he was already fitting rings from largest to smallest on his own. When he made a mistake, he immediately corrected it, without anyone saying anything. I said to him, 'Samuel, fit the blue ring', and he fit it. I looked at Felipe and said, 'My God, it must be all in our heads'", reports the mother [22].

At 2 years old, the boy from Votorantim (SP-Br) already knew how to read and do simple calculations. According to Camila, at just 1 year old, Samuel already demonstrated the ability to form syllables. Currently, he also understands numbers and knows the names of the planets. "It was one surprise after another. He learned names of the planets, how to count to 30 and how to count down from 20 onwards, all by himself". "Today he knows how to count to 100", says Camila [22].

Samuel's family also noticed that the boy showed an interest in the English language and mathematical calculations, which he learned by watching educational cartoons. "With the calculations, he would come to me and say: 'Mom, 2 + 2 + 2 is 6 and 2 + 2 + 2 +2 is 8'. It was then that I started doing some math with him", she explains.

According to educational psychologist Elisete Beranger, clinical monitoring is important when a child shows signs of accelerated development, as in Samuel's case. "Ideally, the child should be assisted by a psychologist so that, initially, through anamnesis, hypotheses can be identified as to why the child is so easy to learn. From there, the little one can be subjected to an IQ test, to identify areas of development that may or may not be above their age range. Only then will it be possible to diagnose whether the child has an IQ above average, high or even prone to an autism spectrum disorder (ASD)".

Samuel's parents said that they recently acquired medical insurance and, based on this, they intend to accompany their son with psychological consultations and take IQ tests. For them, it is impossible to control the feeling of pride when seeing their child display so much knowledge at just 2 years old. "I feel privileged that the Lord gave me Samuel this gift, because this is certainly a gift from God. I'm very proud", says the mother. "I feel proud and privileged to have a child like Samuel", concludes the father [22].

10.2.11   Constantine and the Catholic Church

Emperor Constantine the Great or Constantine I, The Great, was born in 274 and died in 337; he was Emperor of Rome from 306 to 337, for 31 years. He was the son of Constantius Chlorus and Helena, a Christian who became Saint Helena. He married Faustina, daughter of Maximilian Hercules. In the year 313, this emperor gave freedom of worship to Christians and, from then on, Christianity began to add new followers in Rome, becoming the official religion of the Roman Empire in 390, an act instituted by Emperor Theodosius. With this, Christianity was already spreading throughout almost the entire known world, even penetrating the noble class. The emperors tried by all means to persecute Christians with the power of arms, but had little success, as they believed in something stronger than the Roman Empire: God [23–25].

After the death of Emperor Galerius, power was divided between Constantine and Maxentius, who called himself emperor, but the soldiers acclaimed Constantine as emperor. They both aspired to absolute power. This dispute ended on 28 October 312, with Constantine's victory at the Milvian Bridge. Christians at that time said that Constantine saw a cross in the sky with the Latin inscription *In hoc signo vinces* ('with this symbol you will win'). However, his conversion to Christianity only took place after being baptized in 337, shortly before his death.

The Edict of Milan in 313 gave freedom of worship to Christians stating, "We would do well to completely annul all the restrictions contained in previous decrees concerning Christians, restrictions that are odious and unworthy of our clemency, and to give complete freedom to those who wish to practice Christ's religion". The Pope on this occasion was Miltiades, who was the 32nd Pope of the Catholic Church, who became Saint Miltiades. Therefore, there is no need to say that Constantine was the founder of the Catholic Church, since he only gave freedom to Christians, ending two and a half centuries of persecution and intolerance [25].

10.2.12   Testimonies of Sensitivity to Subconscious Information

On the night of 26 January 2013, around 800 students from the Veterinary Medicine, Pedagogy, Agronomy and Zootechnics courses at the Federal University of Santa Maria-RS went to celebrate a party at Night Club Kiss (Boate Kiss), which was then destroyed by fire. This night club was located at Rua dos Andradas, 1925 (see Figure 10.1), in the center of the city of Santa Maria, which is 300 kilometers from Porto Alegre, in Rio Grande do Sul-BR. The party would be marked by the performance of two bands. At around 2:30 am on 27 January 2013, it was the turn of the musical group Gurizada Fandangueira to take the stage. In addition to the songs, the presentation promised to entertain the students with a pyrotechnic show. This was the cause of a fatal fire at Kiss Nightclub, which celebrated its 10th anniversary at the beginning of 2023. Two hundred and forty-two people died in the tragedy and more than 630 were injured, in addition to the marks left throughout the municipality. This tragedy marked the lives and stories of people who lost a friend or family member. Certainly, the case impacted the entire city of Santa Maria, in the interior of Rio Grande do Sul [26].

It is difficult to find someone in Santa Maria who does not have something to tell about what they were doing that fateful night or how the city changed after the tragedy. This is the case of Nicollas Antunes, who was 12 years old at the time. He said, "I was a kid, I was little, I don't have many memories, but the main one, what they

FIGURE 10.1   Front of Kiss nightclub after the fire.

tell me, is that my mother was supposed to be at the club that day. At the last minute, they decided not to go to the club and go to a show closer to our house. That day, my mother lost two friends".

Resident Lucy Polidoro Paim says her son went to the club that night and survived. Even with the relief of having saved her son, she feels the pain of the loss of other families. "We lived a year of profound sadness. Every day that passed, when I passed by Kiss, I relived everything. It's an inexplicable thing, only those who lived it will know", she said.

The retired professor from the Federal University of Santa Maria, Maria de Lourdes Pippi, lived near the nightclub. She says that sadness has taken over the city. "The city was really sad. There was no way to forget it. Everywhere we went there was a mother, a father or someone suffering. They did it here in the square, they still have it, a tent where they put all the photos and they stayed here".

A psychologist from Eixo Kiss, from the Santa Maria psychoanalysis collective, Vanessa Solis Pereira, says that this collective feeling comes from a sense of belonging. "It has this dimension that it is with a neighbor, with an acquaintance. Most people in the city have lost someone in the most direct sense, and we work with the idea that we all have" [26,27].

A detailed, personal and real testimony about the Kiss nightclub tragedy was told by Lúcia Silva (fictitious name) relating to her son José Silva (fictitious name), aged 19 at the time, concerning what happened to them on the date of the tragedy. Lúcia says, "Why didn't I let José, my son, go to Kiss Nightclub that night? He was quiet at home and his friends started calling him asking to go to the club together. José started asking me for money for this and I refused. Therefore, I said I would not allow him to go to the nightclub because on Thursday he had already gone with his friends who did everything that could have been a party. Why would they have to go to the nightclub again on Saturday? Then I told him that I did not have the money for this. Why I did not want him to go, I do not know. It was something different he gave me when he asked me to go to the club. Then, I sat on the sofa, he sat on the computer, and he kept asking me, 'Mom, can you give me money because I'm going, you're miserable, you are stingy' and a bunch of other things and I said: 'You're not going!' Then, I sat on a sofa with him and we started arguing and he was always disrespecting me; and I said I was not going to give him any money. He continued to call me a cheapskate, miserable, everything else ... but that he wouldn't go to the club that night".

Lúcia then continues her story, saying that it was one o'clock in the morning, close to two o'clock and her friends were calling her son, insisting on him and saying that they were already in line to enter the club. Then he turned to me and said, 'Mom, aren't you going to make up your mind?' In addition, I repeated: 'I'm not going to give you money!' Then it was about 15 minutes before two in the morning, and another group of friends called from Camobi, district of Santa Maria-RS, saying that they were in a small car and if he would lend mother Lúcia's car for them to go, because the car was bigger. To which I said, 'José, I'm absolutely not going to lend you the car, because if you leave here and go to the club you already know what you're going to get'. To which he replied: 'Mom, I'm not going anymore, because now everyone has gone in and they're already celebrating and there's no way to get in anymore!'.

"Then the friends came from Camobi, parked the car in front of the building where we live. Therefore, I decided to lend him the car key so he could pick up his friends who went to an avenue known as Presidente Vargas, close to Patronato. I continued drinking mate (tea) because I was no longer sleepy. When José's friends arrived at Patronato and then José called me saying, 'Mom, you do not know what happened: the club is on fire!' To which I said: 'For God's sake, you don't go in the queue by car because the car has gasoline, it's dangerous, and we don't know how much fire there is; you're not going in to look for anyone! You're not going even close to the club' since I did not know the extent of the fire, whether it was just inside the club or whether it was spread out across the street. I was very anxious about this and told him not to go in because it would be very dangerous. He went with his friends towards the nightclub and left his cell phone on saying: 'Mom, I'm not near the club!' Then he told me that they were already removing dead people and bodies from inside the club. I asked him, for God's sake, not to go in to look for his friends. He didn't come in and I asked him to leave his cell phone on at all times.

"From then on, he and his friends who had not gone to the club went to all the hospitals in Santa Maria to look for friends who had gone to the club. They did not find them because they were already dead and they did not know it. In one of the hospitals they entered, there was a girl, who was at the nightclub, and she begged them to save her saying: 'Please don't let me die!' They had nothing to do. That's what my son said".

Lúcia continues, "My son then arrived home at 6 am. Then I said, 'You saw, my son, why didn't I give you money?' He says, 'Mother, we'll talk later'. I said, 'No talking later'. I had not slept all night and was waiting for him, with my cell phone always in my hand. Strangely, he arrived very angry where I was sitting when he said to me, 'Mom, I want to ask you a question! Why did your God kill my friends?' Then I said, 'You calm down and I'll explain why!' One of the questions I asked at that time was 'Did you see why your mother did not give you money, why me too, I do not know! It was an intuition that gave me, something that I don't know what it was, I can't explain it! I didn't give you money because you could be one of them, have you thought about that?' He returned home to get money and go to another city where the funeral of one of his friends who had died on that occasion would take place" [25,26].

10.2.13 SACRED HAND GESTURES – THE MUDRAS

The body can heal itself on its own, using a complete map of the energy that runs through the human body representing emotions or physical behaviors. This has its own science known as Mudra, which is a Buddhist and Hindu discipline that has been used for thousands of years due to its great benefits. Ancient Buddhist masters believed that each finger on the hand has an internal relationship with some type of physical and/or emotional element of the body itself. Although this truth has been kept secret for many years, in a short time the incredible benefits that Mudras have provided have been revealed. Therefore, they believe that it is enough to practice them once a day [27–29].

The text below describes how to use each Mudra and what benefits they can provide. Some unique ways to enhance the effects on the body and mind are also explained. The origin of Mudras dates back many years ago. It is not known exactly where they came from, but it is estimated that they are so old that they date back to certain rituals even before the Christianization of the Nordic peoples. Multiple religions have adapted these gestures and it is possible to see some adaptations of this practice in Catholic, Muslim and many other teachings. It is worth noting that they are not directly linked to a particular religion, as Mudras go beyond a mystical expression and are closely related to a medicinal and therapeutic approach derived from the practice of yoga and meditation.

To explain it more simply, Mudras have the ability to heal people in all possible ways, as they encompass a wide variety of healing forms in a single practice. Among the categories that Mudras cover, the practice of *ayurveda* ('ayur' is life and 'veda' is knowledge) stands out, which comprises compression and meridians. Each of them offers a unique contribution. The practice of *ayurveda* is an Indian art that considers that each disease is an imbalance in the body that needs help to regain its natural rhythm. In other words, the brain creates the disease and everyone's responsibility is to balance this affected element [27–29].

The case of body compression is a Western technique that will perform stimulation using specific fingers for several minutes to obtain positive results, mainly in the functioning of internal organs. Finally, there is the practice of activating the meridians, which are the channels in the body that help with energy balance and reestablishing harmony in the body, responsible for breathing, digestion and the functioning of some specific organs.

Mudras can be performed at any time, but it is recommended to do them at home during a meditation session as these practices derive from yoga and require concentration to have effects on the body. It is usually recommended to constantly practice these Mudras and combine them with other centuries-old practices, such as yoga or meditation to bring peace to life and health to the body. Here are some examples of Mudras:

- **Kubera Mudra**, dedicated to the god of prosperity, Kubera. For this Mudra, the tips of the thumb, index and middle fingers of both hands are placed together, forming a reference in the form of a hoop, while the remaining two fingers are brought halfway across the palm and left to rest there. This position should be maintained for five or ten minutes, or even longer if necessary. The idea of this Mudra is to meditate while it is performed as this position tends to have a highly spiritual effect, focused on increasing the percentage of energy that is positive in the body, in addition to promoting abundance. Other things that this Mudra can provide are confidence and serenity, mainly because it is in direct contact with each person's meridians, which are the body's Chi energy channels. It is recommended to practice it daily when waking up for approximately five minutes as it will almost immediately make the person feel more energetic and positive and start the day on the right foot.
- **Apan Mudra,** or energy-giving Mudra. This is closely related to the body's energy and to practice it you need to bring your thumb, middle and ring fingers

together and then extend the remaining fingers. This Mudra can easily be done for 10 to 20 minutes in a meditation session or at any time as needed. It is excellent for the body, but everything has a limit, and this technique should not be abused, as it can eliminate toxins from the body that are necessary to maintain an adequate intestinal pH balance. Apan Mudra also influences emotional health, as it is perfect for shaping positive manifestations, helping with emotional balance and making conscious decisions. This is because it is related to the wood element and also contributes to the proper functioning of the practitioner's liver and bladder.

- **Tse Mudra**, the exercise of the three secrets. This is one of the most intense and powerful exercises there is. To perform it, place both hands facing up on your legs and place your thumb at the base of your little finger. Then surround yourself with the remaining four fingers while taking a few deep breaths. During this period, it will be necessary to hold your breath for a few seconds while repeating each person's personal mantra. When exhaling, it is important to open the mind and imagine that all worries, fears and unhappiness are leaving the practitioner's life to give way to happiness with each repetition of the Mudra. You can do it seven times or more after entering a state of deep meditation, remaining completely focused on the activity. According to Chinese monks of great importance, this Mudra wards off sadness, reduces fear, and wards off misfortune and bad luck in relation to health. It helps with depression and mental problems of various origins.

- **Shiva Linga Mudra**, the energy recharger. Place your left hand at the level of your abdomen with the palm facing up. Then, place your right hand over it in the form of a fist, but make sure only the thumb is erect, simulating that you are giving approval to someone. Lastly, keep your elbows pointed outwards. Do this as many times as desired or twice a day for four minutes. The right hand in this mode signifies the masculine strength of the phallus, as the shape of the hands and the erect thumb in the Linga Mudra resembles the erect penis (phallus) of Shiva. Shiva is the destructive and transformative aspect of the highest deity in Hindu mythology, just as the phallus is the symbol of a new beginning. Although it may seem negative, Shiva can be considered as someone who helps control each person's life and respect the cycles of life. Although it is sad for flowers to wither, this process is necessary so that new life can be generated. Everything is a cycle, as everything has a beginning and an end, just like the stages of life and the internal interaction we have with emotions. This Mudra is related to the water element, which is present on the outer edge of the hand and in its palm. In this exercise, the thumb represents the flow of energy that will pass through the lungs and head, traveling throughout the body and filling it with new refreshing sensations. It can be very beneficial against tiredness and dissatisfaction resulting from depression or constant physical exertion.

- **Shakti Mudra**, in honor of the goddess of vital energy. Extend your hands and one of your ring fingers with your pinkies while closing your other fingers over your thumbs, which should be placed inside the palm of your hand. Concentrate your breathing in the pelvic area and slow your exhalation a little.

You can perform this exercise three times a day for 12 minutes, as necessary. This is one of the most physically beneficial Mudras and helps with the respiratory drive of the inner chest region. It also helps people with anxiety and insomnia to fall asleep easily in addition to helping women control menstrual cycle pain.

- **Uttarabodhi Mudra**, divine illumination. Uttarabodhi is a compound Sanskrit term, where *uttara* refers to 'upward' and *bodhi* refers to 'closer to awakening'. For this exercise, you must cross your hands in front of the solar plexus at stomach level. Then, place your index finger and thumb on top of each other, pointing them up toward the ceiling and your index fingers down toward your stomach. It is important to highlight that you can practice Uttarabodhi anywhere, at any time and for as long as you wish, however it is especially recommended when the person feels physically and mentally tired or stressed by everyday situations. This exercise provides its benefits through its connection with metal and has the greatest impact on the nervous system. That allows the body to relate it both to the surrounding environment and to the energy coming from the Universe. Such energy is well known as Chi. Many artists use this technique to simulate inspiration and creativity, as it promotes energy in relation to physical health, helps improve the functioning of the large intestine and allows the heart to pump blood in a healthier way to other parts of the body.

- **Ksepana Mudra**, the gesture to understand and release. For this last Mudra, the index fingers are placed flat against each other while the other fingers are brought together at the same time as their tips rest on the back of the hands. When you cross your thumbs and place them in each other's space, you will notice that there will be a small space between them. Breathing deeply and holding this gesture for a few breaths, about 15 minutes, will be enough, trying to concentrate on the exhalation. When finished, place your hands on your thighs with your palms facing up. Although it seems a little complicated to do this, with practice it is easily achieved and brings global benefits. This exercise stimulates the elimination of toxins through the large intestine and helps clean the pores. It is ideal for removing negative energy, but it should not last for more than 15 breaths, which could cause the new energy to dissipate.

10.2.14 SPIRITUALIST LEADER IN BRAZIL

Born in the interior of Minas Gerais-BR, in 1924, Isabel Salomão de Campos is from the first generation of Brazilians descendent of a family of Lebanese immigrants. Raised on a farm in the backlands of the state of Minas Gerais, central Brazil, she showed that she was different early on, when at the age of 9 she saw and heard things that she could not explain, and she blessed people without access to medicine and medical care. At 14, she alone obtained authorization from the mayor of her town to create a school for the children of the settlers where she would be the teacher [30].

It was at the beginning of her adult life that Isabel had a broader understanding of her vocation when she discovered that the 'things' she had interpreted since childhood were 'spirits' communicating with her. Since then, Isabel began a long process of learning about spiritualism, being the first woman to raise her voice publicly to speak about this invisible world in Brazil. The fight against religious prejudice marked her life and the invisibility imposed on women. Obstinate, she created two other schools and removed more than 500 children from the streets throughout her life. She also built a solidarity network that serves families in situations of social vulnerability in more than 40 neighborhoods in the city of Juiz de Fora-MG, where Isabel still lives and where she founded Casa do Caminho, a center not only for celebrating her faith, but also for welcoming people.

As an adult, Isabel began to study spiritualism even more, becoming the first woman to speak openly about this subject in Brazil. At 22 years old, she was already the first female voice to speak publicly about spiritualism, when the belief was not even considered a religion in the country, shortly after the end of the Second World War. Amid prejudice and criticism, the healer created two other schools.

In her career, she was a teacher at a school that she founded for the children of farm employees in the backlands of the state of Minas Gerais. For Dona Isabel, what most marked her journey was interpreting the beauty with which the world of spirits works for the benefit of humanity. At 100 years old, now dead, Isabel Salomão de Campos led her peaceful life next to A Casa do Caminho, which she helped build. Literally, as the book *The Two Worlds of Isabel* narrates, this woman even carried cans of water on her head to help with the construction of this house and the entire time she was involved in the activities carried out there [30].

10.3   SCIENCE EXPLAINS SUPERNATURAL PHENOMENA

If a person admits to having had a supernatural experience, there could be three reasons: 1) perhaps they should admit this experience; 2) maybe this person had something deep that he/she wanted to share with other people; 3) both. Most people have trouble understanding this. Studies show that this ethereal experience can be explained using scientific models of the mind and body to make a relationship between them. The largest study on this was carried out in 1894 at a British university called the Society for Psychic Research (SPR), which published a survey on hallucinations involving more than 17,000 people from the UK, the USA and the European continent. The research aimed to find out how common it was for people to receive apparently impossible 'visits', with harbingers of death. SPR felt that these experiences were too common to be random, 1 in 43 people in this study. Among the SPR's patrons were former British Prime Minister William Ewart Gladstone and the Poet Laureate, Alfred, Lord Tennyson [31,32].

Many of the cases collected by the SPR appear to be hypnagogic cases, which are hallucinatory experiences that occur at the edge of sleep. Studies have suggested that several religious experiences recorded in the 19th century were linked to hypnagogy. Presences have a particularly strong relationship with sleep paralysis, which affects around 7% of adults at least once throughout their lives. In sleep paralysis, the muscles remain frozen as a remnant of rapid eye movement (REM) sleep, but the mind remains active and awake. Studies have shown that more than 50% of people with sleep paralysis report encountering some presence. The Victorian-era presences documented by the SPR were often benign or comforting. In modern examples, however, the presences caused by sleep paralysis often emanate evil.

Societies around the world have reported stories of nocturnal presences. Examples range from the 'friar with the hole in his hand' in Portugal, who managed to infiltrate people's dreams, to Ogun Oru. His victims were believed to have been bewitched. One wonders, then, why experiences such as sleep paralysis would create a sense of presence. Some of these researchers have focused on the specific characteristics of the person waking up in this unusual situation.

Most people find sleep paralysis a frightening experience even without the hallucinations. In 2007, some sleep researchers put forward the idea that if people woke up vulnerable and paralyzed, this would make them feel threatened in their instincts and the mind would be filling in the gaps. If a person felt like prey, that meant there would be a predator. Another approach was to observe the common characteristics between visits and other types of presence during sleep paralysis. Research has shown over the last 28 years that presences are not just frequent in the hypnagogical scenario. They have been reported also in cases of Parkinson's disease, psychoses, near-death experiences and bereavement. These findings indicate that it is unlikely that this is a sleep-specific phenomenon [31,32].

A series of experiments demonstrated in 2014 that undoing expectations appears to induce a sense of presence in healthy people. In these experiments, researchers tricked participants into feeling as if they were touching their own back by synchronizing their movements with those of a robot directly behind them. The brain perceives this synchronization, deducing that the person is producing that sensation. When the synchronization is interrupted, causing the robot's touch to be slightly out of tune, people may suddenly feel that another person is present, that is, a ghost in the machine that is an air of expectation of something similar to a hallucination [32]

The logic of undoing expectations seems to induce a sense of presence in healthy people that can also be applied to sleep paralysis. All the usual information about the body and the senses is destabilized in this context so that the feeling that there is another person there with the person is not surprising. It may feel like it is another presence, but in reality, it is the person themselves.

In a 2022 research, an attempt was made to track singularities between the presences verified in chemical data, spiritual practices and pockets of resistance, all known to produce certain hallucinatory phenomena, including presences. In all of these situations, several aspects of the presence sense were quite similar. The patient felt, for example, that the presence was directly behind him. The three groups described presences related to the dream, but also presences caused by emotional factors such as grief and loss. Despite having lost the presence of the specter, the loss of the felt presence has actually just begun. Scientific research may provide comprehensive explanations or may require several theories to clarify all these cases of presences, but the encounters described in the *Phantasms of the Living* are not echoes of a bygone era. People have disturbing experiences, and if they have not, they probably know someone who has [31,32].

10.4   CONCLUSIONS

The observations contained in this book are based on the interpretation that can occur with an organ as complex in form and influenced by the environment as the human brain. The enormous number of neurons and the chemical or electrical connections contained therein can suffer many external interferences from distinct sources, leading a person to interpret things based practically on nothing, that is, on what they hear, feel and see around them or in their imagination. Cell phones, radio and television stations, exposure to the Sun, means of communication, the movement of metallic masses in cars and means of transport, human relationships through contacts, fights, all of these and much more, can contribute to alter the behavior of the human imagination and the mind.

All the cases reported in this chapter clearly seem to be preliminary subconscious realizations of what could happen. In other words, a gathering of subconscious information comprising facts and sensations that could involve the premonitions that occurred, in a more or less clear way. The exercises suggested by the texts seem to be nothing more than helping to sharpen the subconscious to understand what may happen in the coming period, especially in cases where there is a very direct relationship between the seer and those involved.

It is very difficult to say anything about the extrasensory manifestations of human beings due to their complexity and people's experiences in a world so full of ignorance, beliefs, irradiations and religions. These extrasensory manifestations are considered to be abnormal facts that are already historically rooted in human beliefs and religions due to either ignorance, interests, traumas or complex inexplicable fears. However, it seems that brain manifestations are still not sufficiently known or explicable due to their complexity and perhaps hampered by the amount of beliefs that have existed since the beginning of humanity. There were times when those who did not follow in the beliefs or interests of the most powerful were condemned to death, expelled from religions, exiled, silenced in some way or, at the very least, socially segregated.

It is known today that there are so many frequencies, natural or artificial, crossing the Universe that cannot be consciously or sensorially felt, but that exist and definitely have an influence on people's behavior. How could one explain that the skin can only feel from 0–200 vibrations or cycles per second (Hz), the ears can hear from 20–20,000 Hz and the eyes up to 790×10^{12} Hz and that there are no other detecting organs in the sensitive body for so many other frequencies in the Universe? Why could only these well-known frequencies affect human consciousness and beliefs? This diversity includes the frequencies of microwave ovens, which are 2450×10^6 Hz, radio waves ranging from 3 kHz to 300 GHz (1 GHz = 10^9 Hz), those of the X-ray that range from 3×10^{16} Hz up to 3×10^{19} Hz, gamma rays ranging from 10^{19} Hz up to 10^{24} Hz, and many others, including human beings' mind irradiations and receptions. Of course, the effects caused by all of them depend on the intensity with which they are generated at the source, the sensitivity of each person and how they can be received by living beings.

From everything explored in this book, it can be concluded that human beings still need to study a great deal so that they may have an acceptable maturity, understanding and knowledge about the effects of vibrations or frequencies on the mind, interpersonal behavior, individual behavior and bodily communication in general.

REFERENCES

[1]   P. Paixão and C.S. Silva and J.L. Brand, Letargia e Hipnose Sem Magia, 8th ed. Graphics and Publishing Padre Berthier, https://pdfcoffee.com/qdownload/letargia-e-hipnose-sem-magia-3-pdf-free.html, 1995.

[2]   A. Túlio (Ismael Gomes Braga), https://aron-um-espirita.blogspot.com/2013/11/uma-historia-que-se-repete.html, 06 Nov 2013.

[3]   F. Database, Marist brother uses needles to explain "miracles", Company Folha da Manhã Ltda, Campos Eliseos, Folha da Noite, http://almanaque.folha.uol.com.br/cotidiano_01out1959.htm, 01 Oct 1959.

[4]   T. Cicoria, Doctor who became a successful pianist after being struck by lightning, BBC, https://g1.globo.com/mundo/noticia/2020/12/25/o-medico-que-virou-pianista-de-sucesso-apos-ser-atingido-por-um-raio.ghtml, 25 Dec 2020.

[5]   The Illiterate Who Learned 3 Languages While Dreaming. YouTube, Brasil Acadêmico, http://blog.brasilacademico.com/2013/11/o-analfabeto-que-aprendeu-tres-idiomas.html, 2013.

[6]   L.B. D'Angelo, L.A. Ribeiro and R. Bulcão, NotaTerapia, www.youtube.com/watch?v=4NrXdPu-oYM, 2021.

[7]   J. Borba, Galinha Tonta no Domingo Espetacular, https://tonocosmos.com.br/o-menino-que-aprendia-idiomas-enquanto-dormia, 30 Mar 2014.

[8]   O. Sacks, Hallucinations. National Bestseller, Oliver Sacks Foundation, original Easton Press, 06 Nov 2012.

[9]   B. Alderson-Day – The Conversation (republicado sob licença Creative Commons). Universidade de Durham, no Reino Unido, www.bbc.com/portuguese/articles/cv2jweqyg19o, 14 Apr 2023.

[10] J.F. Peres, A. Moreira-Almeida, L. Caixeta, F. Leao, and A. Newberg, Neuroimaging during trance state: A contribution to the study of dissociation, PLoS ONE, 7(11), e49360. https://doi.org/10.1371/journal.pone.0049360, 2012.

[11] D.S. Roggo, Life after Death: The Case for Survival of Bodily Death. Aquarian Press, 1986.

[12] C.A. da Silva, Chico Xavier's Letters: A Semiotic Analysis, Master Dissertation, Advisor: Prof. Dr. Jean Cristtus Portela, São Paulo State University, UNESP, FCLAr, Mar 2012.

[13] M. Marques, Russian Boy Tells Strange Story and Says He Once Lived on Mars, Curiosities Unknown Facts, Images: Freepik Uol, 10 Jan 2023.

[14] Bio Oficial da Temporada 2020–2021. eblackburn@kanzenarts.com and www.gavingeorge.com, Feb 2021.

[15] Catholic Diocese of Dallas, Canonization of Mother Teresa of Kolkata, Pastoral Center Offices, info@cathdal.org, 02 Sep 2016.

[16] M. Loudon, (1 de janeiro de 1996). Review of the missionary position: Mother Teresa in theory and practice, BMJ: British Medical Journal, 312(7022), 64–65.

[17] A. Chatterjee, Mother Teresa: The Final Verdict. Meteor Books, 2003. ISBN 81-88248-00-2, Introduction and first 3 chapters of fourteen (without pictures). Critical examination of Agnes Bojaxhiu's life and work.

[18] BBC – 19/12/2022 – https://g1.globo.com/saude/noticia/2022/12/19/morte-lucida-as-pessoas-com-alto-nivel-de-consciencia-e-sensacoes-extracorporais-quando-o-coracao-deixa-de-bater.ghtml.

[19] A. Sleutjes, A. Moreira-Almeida, B. Greyson, Almost 40 years investigating near-death experiences. An overview of mainstream scientific journals, The Journal of Nervous and Mental Disease, 202, 833–836, 2014.

[20] D. Gayle, Near-death experiences occur when the soul leaves the nervous system and enters the universe, claim two quantum physics experts, Daily Mail (online), 30 Oct 2012.

[21] https://experienciasdequasemorte.blogspot.com/2014/05/blog-post.html.

[22] https://g1.globo.com/sp/sorocaba-jundiai/noticia/2023/04/16/aos-2-anos-menino-prodigio-le-faz-calculo-matematico-e-conhece-o-nome-dos-planetas.ghtml.

[23] B. Bleckmann, Chapter sources for the history of Constantine, in Noel Lenski (ed.), The Cambridge Companion to the Age of Constantine. Cambridge University Press, https://doi.org/10.1017/CCOL0521818389.002, 2005, pp. 14–32.

[24] D.M. Nicol and J.F. Matthews, Constantine I, in Encyclopedia Britannica, www.britannica.com/biography/Constantine-I-Roman-emperor, 16 Aug 2023.

[25] F. Aquino, História da Igreja – Idade Antiga, 1st ed. Editora: Cléofas, ISBN: 978-85-88158-90-0, 2015, 440 pp.

[26] G. Brum, Kiss nightclub: How did residents of Santa Maria deal with the tragedy? Agência Brasil, Rádio Nacional, https://agenciabrasil.ebc.com.br/geral/noticia/2023-01/boate-kiss-como-moradores-de-santa-maria-lidaram-com-tragedia, 27 Jan 2023.

[27] F. Previdelli, 10 years ago, the tragic fire at the Kiss nightclub shook Brazil, Aventuras na História, UOL, https://aventurasnahistoria.uol.com.br/noticias/reportagem/incendio-da-boate-kiss-completa-sete-anos-sem-o-julgamento-dos-reus-e-o-pagamento-de-indenizacao-para-vitimas-e-familiares.phtml, 27 Jan 2021.

[28] G. Devi, Esoteric Mudras of Japan. International. Academy of Indian Culture & Aditya Prakashan, ISBN: 9788186471562, 1999.

[29] Semple D., Smyth R., Chapter 1: Thinking about psychiatry, in Oxford Handbook of Psychiatry, 4th ed. Oxford University Press. ISBN 978-0-19-879555-1. https://doi.org/10.1093/med/9780198795551.003.0001. These pseudoscientific theories may confuse metaphysical with empirical claims (e.g. Ayurvedic medicine), 2019, p. 24.

[30] D. Arbex, Isabel's Two Worlds. Editora Intrínseca Ltda, ISBN: 978-85-510-0658-0, 2020.

[31] Statistics Report. Sri Lanka Institute of Indigenous Medicine. Nov 2011, archived from the original on 24 Apr 2012.

[32] Arham Dilshad, What science can tell us about the experience of unexplainable presence, BST, https://theconversation.com/what-science-can-tell-us-about-the-experience-of-unexplainable-presence-201323, 26 Nov 2024.

Index